Jack St. Clair Kilby

A Man of Few Words

Texas Instruments Annual Report 1959, page 10

"a. *An engineer displays a revolutionary new product, the Solid Circuit semiconductor, TI's most significant electronic development since the silicon transistor and the greatest advance achieved to date in micro-miniaturization of electronic equipment."* An engineer?

Jack St. Clair Kilby

A Man of Few Words

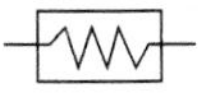

a brief biography
by

Ed Millis

2008

ISBN 0-9718402-8-8

Edwin Graham Millis
9405 Forestridge Drive
Dallas, Texas 75238

Other books by Ed:

> ***High Voltage, Gunpowder and Mousetraps****, or the Nearly Perfect Childhood of Ed Millis*—the story of growing up in Dallas on the "M" streets. 1999 [out of print]
>
> ***TI, the Transistor, and Me****, or My Dis-integrated Circuit Through Texas Instruments*—the lighter side of 37 years at TI, from before the transistor to the 64-Meg DRAM. 2000
>
> ***One Way to Write your Personal Story***—a step-by-step book for the unlearned who really, really want to write and publish their own memoirs, written by one who did. 2002

Cover art by Bev Haskin

First Edition – July 2008

Manufactured in the United States of America

from the little print shop on forestridge

Dedication

THIS BOOK, the life of a creative person, is dedicated to all those who create, whether it be science, letters, art, or simply joy in someone's heart.

Foreword

THE IDEA FOR A BOOK about Jack Kilby grew out of the process of making plans for the 50th Anniversary of his invention, the Integrated Circuit. While much has been written as chapters or sections in books and journals about his invention of the chip and its significance to the world, little is found about Jack's early years, the war years, or his interests outside of engineering. As the idea for a short biography took shape, it became clear that the person who could write Jack's story had to be not only someone who knew him well and had access to archival material as well as contacts with family and friends, but someone who could paint a broader picture of Jack. Jack was an engineer. What we needed was an engineer who could write and do it fast. Only one person met the specs, Edwin Graham Millis, or simply, Ed—part engineer, part detective, writer, story teller, former motorcycle rider, and a long-time friend of Jack Kilby.

But why a book about Jack Kilby? If for no other reason, it's a great human story about a lanky kid from Kansas whose college years were interrupted by World War II, who came back, as did millions of other veterans, to a much different world. It's the story of one quiet, unassuming engineer who went on to solve one of the most vexing problems in electronics in the 1950s; one that had eluded the research staffs of some of the world's largest corporations and most prestigious universities. The story of his improbable invention of the IC at a small upstart company in Dallas would itself make an interesting commentary on the process of innovation. But none of these themes alone led to this book.

In late 2005, SMU announced that Texas Instruments was donating its historical archives to its DeGolyer Library, and also that the Kilby family was donating many of Jack's awards, books, documents, and personal items to SMU. This included thousands of photographs and negatives taken by Jack in his lifetime. In going through the Kilby collection, Dr. Anne Peterson, Curator of Photographs for the DeGolyer Library, recognized the quality and artistry of the photographs. She took some of them to the Meadows Museum, an art museum on the campus at SMU, and they were struck by the scope and artistry of Jack's work. Out of this, with the help of Texas Instruments, grew the plan to exhibit around 50 of his works at the museum, along with a display of some of his inventions. Jack's closest friends knew about his interest in photography, but few others did. As Jack's photographs and papers were examined, it became clear that there was much more to the story. Ed Millis has researched the genesis of Jack's interest and passion for photography. In this book, he will tell that story, and we will see the character and personality of a man whose life has affected all of us. Jack showed us what it meant to be a real engineer—to be a problem solver—and to do it with integrity.

In the year 2000, Ed Millis published a slightly irreverent book documenting his 37-year career with Texas Instruments, *TI, the Transistor, and Me.* On the back cover of his book appears one of the shortest book reviews in history from his friend, Jack Kilby. In typical fashion he simply said, "I enjoyed it." Like Kilby, Ed believes in the economy of words and has a knack for making the complex simple. At times, he can't resist the urge to comment on society in general, or puncture a balloon or two. At a time in life when most retirees might be thinking of slowing down or taking it easy, Ed had the courage to take on the challenge of researching, writing and producing a biography on a tight schedule. When asked if he would do the Kilby book, he didn't flinch. He said, "I'll do it. He was my friend."

Max Post, TI Retiree
May 1, 2008

Preface

THIS WAS GOING TO BE THE YEAR for me to slow down and kick back a little. Selfish as it may sound, I wanted to get started on some of my own projects. I passed up an opportunity or two to assist in useful efforts for others and found that, with practice, I could say "no" and make it stick. I had a list of things I'd been wanting and needing to do, like finish cataloging family papers, redoing the bookshelves in the den, and sorting my books, and, as Snuffy Smith used to say, time's a-wastin'.

It was going to be a great year. January brought the 30% off sale of Elfa shelving, and by the end of January I'd finished hanging a lot of new shelves and was testing some really cool lighting in the room. There's nothing more exciting to a book collector than empty shelves.

The only outside activity that I'd continued was working with the small group of Texas Instruments retirees that Max Post, hyper-active TI retiree himself, had recruited in the spring of 2004 to help write TI's 75th anniversary book, *Engineering the World.* Max was so pleased with our work on the book that he strained mightily to find more projects for our splendid team, like triaging the contents of 1,500 document storage boxes in the TI archives, and then sending them to SMU for management. And, oh yeah, we need to do the same thing for the TI artifacts collection—another 1,500 acquisitions for SMU. Then, just as we were running out of work and threatening to disband, Dr. Anne Peterson of SMU's DeGolyer Library came to

TI in late January with a great idea—have a jointly sponsored dual exhibit of Jack Kilby's photographic art and his scientific side. Why not? And so it became "*The Eye of Genius*," to be displayed in the elegant Meadows Museum on the SMU campus, July 12 through September 21, 2008.

Ideas were kicked around for other activities relating to this event, and one suggestion was to have a brief biography of Kilby that could be sold in the Meadow's museum shop during the 10-week show. When I heard the discussion that a book was needing to be writ, I could feel my blood pressure rising. I had begun my secondary semi-career of writing the day after I retired from my engineering career on August 31, 1999, and haven't stopped yet. I truly love to write, usually to the detriment of everything else in my life. Max and I talked about my doing a small brochure but, remembering my 2008 plans, I stood by my guns—this was the year to say "no." Then I thought about it—critical thinking I believe it's called. My critical thinking drowned out any preconceived biases I had about 2008. How could I *not* do a book about Kilby? It was truly the chance of a lifetime. Whether this book sells or not is of no great consequence. I will have written a book for the Kilby family, and for Kilby's great family of friends and, most of all, for my friend Jack. I can't think of a better gift to share. I hope you like it.

Acknowledgements

It's hard to put together a biography of Jack Kilby for two reasons. First, it never occurred to Jack to talk about things in the past, which is the basis for a bio, and consequently events from his early years are slim pickings. Secondly, his later years, after his invention of the microchip, are over-documented. Not that it's a bad thing, since this event in Jack's life is roughly on the same level as the discovery of fire, but it's also difficult to write what has been written so many times. What could I possibly say that would add an iota, or more appropriately, a silicon atom, to what's already out there?

Having said that, here follows my brief story of the life of Jack St. Clair Kilby. I found and used a lot of sources, but I found none more valuable than that of Pamela McConnell Karnavas. She knew Jack well as a friend, not as an inventor or electrical engineer, and had the idea of writing a children's book based on Jack's life. Pam interviewed Jack and his sister Jane Kilby at length in 1989 and 1990 to gather biographical material. For various reasons, the book never was written, and, when I approached Pam about this unique store of fresh information, she immediately and enthusiastically agreed to share it with me. I can't imagine how the early life of Jack and his family could have been written without Pam's unselfish gift.

For Jack's later years, I called upon a variety of friends for assistance, frequently several times:

Phil Bogan
Charley Clough
Harvey Cragon
John Fish
Pete Johnson
Steve and Margaret Karnavas
Kevin McGarity
Jerry Merryman
Charles Phipps
Max Post
Walt Runyan
Len Wetterau
Ken Zapp

Texas Instruments has been very generous in their assistance to me with information and pictures, and permission for their use in this book. I'm especially pleased with the new high-resolution scans of Kilby's famous notebook by Cindy Sheets of TI. Phil Bogan, worker of magic at TI, with great good humor worked his magic for me to make a clear and prompt path to publication.

The help of Dr. Anne Peterson, Curator of Photographs of the DeGolyer Library at Southern Methodist University, was invaluable. In addition to the Kilby photographs, she is the keeper of Jack's archived material, without which much of this book could not have been written. The papers in Jack's files from his military years and from his work at Centralab in Milwaukee were especially valuable. It was like finding a telescope into the past.

And of course, Jack's daughters, Ann Kilby and Janet St. Clair Kilby helped immensely. They both cheerfully shared their stories of growing up in the Kilby household, and lent a different and fresh viewpoint of their mother and father. My many thanks! It's been a pleasure to work with you on this effort.

I have been blessed by the assistance of two able editors: my dear Shirley Remnant Sloat, and my extraordinary friend Max Post. I like you, too, Max, just not as much. Shirley is known among our writing friends, as "DRP," or "Red," short for Dreaded Red Pencil. My late engineer and writer friend Jeff Campbell who Shirley had nursed through two books, never failed to end his emails, "Tell Red hi for me." She has suffered through untold pages of my writings for years, and pumped gallons of red ink on it, much to its betterment. Shirley can do it all, but her editing specialty is untangling my sentences and strange expressions. It always brings a smile to my face to enter her suggestions, and then see, as if by magic, what I really wanted to say all along, and now as smooth as new-mown silk. The grammatical knot had been untied. Shirley, stick around. I need you. A lot.

Max was the leader of the monumental effort of herding 14 unruly retired Texas Instruments employees to do TI's 75th anniversary book, and anyone who could lead that gang could manhandle me without raising a sweat. But Max doesn't manhandle—Max is the consummate gentleman and, in the politest way possible, points out to me the errors of my grammar, punctuation, and speling. But he has found his special editing niche in my works—ferreting out the gaps and missing tidbits in the prose of my story, which, like a rough highway, makes a smoother journey when filled in. By the way, he bit on the "speling" trap that I set above, and marked it "spelling" in the margin with his red pen. Gotcha, Max.

Max also single-handedly tracked down all the information on the Kilby patents, which, I assure you, was not a trivial task. He gets full credit for the information in "Patents." And, if there are any errors, he did those, too.

Bev Haskin, aka Beverly Diane Millis Haskin, is both my daughter and the artist on this book. Her talents are exceptional and her assistance generous. She has been my savior many times with her skilled and critical eye. She and Shirley have worked together in their editings of my works over the past many years, and have invented a new editing mark to add to the usual hieroglyphs we love so well—this new squiggle denotes a phrase or passage of my writing that requires the symbolic sticking of one's finger down one's throat and gagging. Thanks, my dears. I needed that.

And, of course, there are those friends of mine who helped me during the frantic course of putting this book together whom I've left out. I'm sorry, but my brain, you know…

Contents

Chapter One

In the beginning

Hubert and Vina meet, get married, and beget Jack and Jane. They move from Jefferson City, Missouri, to Salina, Kansas, and then to Great Bend, Kansas.

HUBERT ST. CLAIR KILBY and Melvina "Vina" Freitag were born and raised in the village of Mackinaw in south central Illinois. In this small town, the story goes, they chanced to meet in the one-room school house, fall in love, and get married.

The Kilby side of this new family came west from Virginia to Illinois in the 1830s. Vina's grandfather Freitag (Freitag was pronounced "Friday," the English meaning of the name), had come to Illinois from Germany. Vina and her family lived in a comfortable cluster of similar families but, in the cramped quarters of tiny Mackinaw, there was an unseen line between those of German extraction and those who were not. It has been hinted that Mother Kilby wasn't at all pleased that one of her two sons married a German girl, and so the two grandmothers never met.

It was not a common occurrence to earn a college degree in the early decades of the 20th century, but both Hubert and Vina attended and graduated from the University of Illinois in Urbana. Hubert received a Bachelor of Science degree in Electrical Engineering in 1914, and soon began working at the McKinley System, a public utility in Peoria, Illinois. Vina earned a Bachelor of Science degree in Household Science in 1918 that would prepare her for her first job as a hospital dietitian.

In 1919, Hubert left his job in Peoria to serve a short hitch in the U.S. Navy as an ensign, and returned to a new job in Jefferson City, Missouri. There he managed the electrical, artificial gas, street railway, motor bus, and toll bridge properties. Hubert was showing considerable management, or maybe juggling, skill at an early age.

When Hubert and Vina married in 1920, only Hubert's brother and wife were present for the civil ceremony. The rest of the families stayed away because Hubert had stepped over the ethnic boundary of Mackinaw to marry Vina. The wedding was an event of considerable gossip, since anything this daring rarely occurred in Mackinaw. But both Hubert and Vina knew what they were doing, and proved it with a long and happy marriage and two fine children.

Their first child, a son, was born while the couple was living in Jefferson City, Missouri. Jack St. Clair Kilby arrived on the 8th of November, 1923, and was announced on the front page of the November 9 issue of the Jefferson City Capital News. It's worth noting that Jack is in a very select group of people who made the front page of newspapers both coming into and leaving this place. Jack came in quietly and left quietly, only pausing to change the world.

It can only be assumed at this time that Jack and his sister Jane, who arrived a year and a half later, led a perfectly normal and happy infancy and early childhood in the bustling Missouri capital of Jefferson City. But this urban Eden was soon to change, as their father had received a job offer to be the manager of a small rural power company in Kansas. Hubert had been doing very well as the manager of a local power company in Missouri, and this perhaps had attracted someone's attention. Hubert took the offer. In 1927, the family packed up and moved to Salina, Kansas, and Hubert Kilby became manager of the Kansas Power Company.

Hubert stayed in the utility business. He was promoted to president of Kansas Power Company in 1933, and didn't step down as president until 1945 when it merged with a neighboring company. He stayed on as vice president in charge of construction and operations until 1951. At that time, Hubert and Vina moved to Tulsa, Oklahoma, where he served in various executive positions with the

Public Service Company of Oklahoma. He retired in 1959 after 45 years of service in public utility management.

Tracking the Kilby path through the utilities includes another member of the family. Jack's "little sister" Jane followed in their father's electrical footsteps. While living in Tulsa she worked for the Public Service Company of Oklahoma as advertising manager and a director of corporate communications. This was followed by Jane's move to Dallas and employment at Central and South West Services, Incorporated, a public utilities holding company. She served as director of business development and coordinator of marketing and economic development programs in the four CSW electric operating companies. As a grand finale at CSW before her retirement in 1990, she wrote the information-packed book, *The Cactus Patch and How It Grew: A History of the Central and South West System,* the story of the power company.

Jack was four years old when the family moved from Jefferson City, Missouri, to Salina, Kansas, headquarters of the Kansas Power Company. The growing electric company served customers "scattered across rural western Kansas." Salina was several steps down in size from Jefferson City, with a population at that time of about 20,000. But with the rows of stately grain elevators and endless fields of wheat, it was a pleasant and attractive Midwestern town. It had been mildly famous during the 1860s as the western terminal of the Smoky Hill Trail. If you wanted to go west on the Smoky Hill Trail, you passed through Salina. Today it is better known as the place to turn left when driving from Dallas to Denver.

Jane Kilby recalled a story about the morning grade school routine in Salina. As she and her big brother Jack prepared to leave for school, books and notebooks belted together, they would step out the door and check the western sky for a too-familiar black band just above the horizon. This was the sign of an incoming dust storm. If an ominous dark smear was visible, Momma would supply her two children with an extra bandana each that they could be wetted down and wrapped across their faces when they walked home for lunch. Shades of Dorothy and Toto.

After seven years of an agreeable and uneventful childhood in Salina, the Kansas Power Company stirred the Kilby family into action once more. This time the company was moving its headquarters to the town of Great Bend, Kansas, a 90-mile hop. Great Bend was another step down in size, perhaps one third the size of Salina but, for two children in grade school, smaller is better. The childhood memories of Jack and Jane Kilby are mostly recollections of the good times they had growing up in Great Bend. This is the town that Jack would later declare to be his hometown.

Great Bend is pretty much dead center in the state of Kansas, and is situated along the banks of a curve in the Arkansas River that gave the town its name. After a rough beginning from the easterners' migration following the Civil War, and a brief period as a cow town with saloons, Texas cowboys, and shoot-'em-ups, it settled into a pleasant regional trade and service center, boosted by the development of the oil industry in the area.

The move to Great Bend was in 1934, the worst of the Great Depression, but Hubert seemed to be depression-proof, in not only having a good-paying job but hanging on as president of a relatively new-fangled company in those cruel times. Although there is not much information about Hubert—no one thought to write the biography of the father of a future Nobel Prize winner—it seems that he was considerably above the national average in both education and plain old savvy. The Kilby family appeared to have been able to live comfortably through the depression, which put them in a small minority in the 1930s.

Chapter Two

Childhood in Great Bend

"I grew up among the industrious descendants of the western settlers of the American Great Plains."
—Jack St. Clair Kilby

THE KILBY FAMILY moved from Salina, Kansas, to a spacious Victorian house in the small town of Great Bend. The house that they rented was known as the Kamerek House, after the original owner and builder, and was on a corner lot at Washington and Forest in a tree-shaded neighborhood. Jack continued his education in the eighth grade of the Great Bend Middle School, and his sister Jane attended the Washington Elementary School, a few short blocks from the house. As it turned out, Mr. and Mrs. Kilby were settling in for the long haul, as they were to stay in this house until 1951, long after Jack and Jane had completed their educations and moved away to begin their own lives.

Life in the Kilby household was probably typical of small-town midwestern U.S.A., but some of the family customs seem quaint by today's standards, or lack of standards. The evening meal, dinner, was a sit-down affair with a white tablecloth and napkins and food brought in from the kitchen. The plates were served by Father Kilby to each family member. This was not just Sunday dinner, or Thanksgiving, but every evening. Conversation during this time was encouraged, but was not to include controversy.

The Kilbys were a pleasant family. Jane reported that she and Jack were never yelled at, and in difficult times the parents were supportive. Jane does remember that her father, the president of Kansas Power Company, was quite concerned that the other president, Franklin Delano Roosevelt, would nationalize all the power companies in the United States, and he'd end up as a government employee. This didn't happen, nor did his fear that the Soviets and communism would come sweeping across the oceans and swallow North America.

The big house came perfectly equipped for the Kilby family—it had a built-in library. The house was crammed with books and magazines belonging to the owner of the house. The basement had stacks and stacks of magazines, mostly *Harper's* and *Scribner's*, some bound in leather, dating back to the late 1800s. Jane described it as "cabinets full from floor to ceiling in the basement." Jack had been especially proud of a paper he wrote in high school with the help of these magazines. The essay detailing Grant's Vicksburg campaign received an A.

Jack loved to read anything, and there was no shortage of fodder of all varieties. There were encyclopedias, travel stories, histories, and biographies for facts, novels for entertainment, and magazines for variety. For keeping up with the rest of the world, the whole family read their mail-order *Time* magazine.

But just in case they ran short of reading material, the family subscribed to the Book of the Month Club. The Club had been started in the 1920s, and had caught on like a wildfire. Even during the depression it maintained high readership. It was one of the Book of the Month Club selections that required a Kilby family council: the selection that would be arriving soon was *Anthony Adverse*, by Hervey Allen. This monster best-selling novel, listed today as 1,224 pages, posed a pair of dissimilar problems. The house rule was that if a person started a book, that person had to finish it before starting another. And the second quandary was whether young Jack should be allowed to read such racy material. It was a popular piece of mildly bawdy fiction, and the book had been offered by the Book of the Month Club to coincide with the release of the well-advertised Hollywood film of the same name. This book contained stories involving illegitimate children, midnight skulking, and occasional bodice-ripping. Jack's father brought up the question as to whether Jack should read this adult-oriented book. The resolution came quickly from Jack's mother: if we purchased this book and it's in our house, Jack should be able to read it. End of discussion. Whether he could get through 1,224 pages would be Jack's problem.

Many years later, Jack was quoted as saying that as the fields of knowledge become narrower with greater specialization, the only way to broaden one's knowledge is to read. *Anthony Adverse* probably did just that for young Jack.

The Kilby family was well-educated and well-read, and not just for that time. They would be considered above-average in today's society also. During an interview, Jane was asked about the family dinner-table conversation. She said she remembered once, when she or Jack had been studying Shakespeare in high school, a scholarly recitation begun by the child was completed by the parents. They knew Shakespeare, too, and weren't afraid to use it.

This remarkably broad-minded and permissive family did have a few rules for the children. Father said that curse words added nothing to the language, and that there were better words to use. Mother added the sage advice that although we may know all that was going on, we didn't always need to talk about it. Good guidance for all people, then and now

Jack took advantage of the freedom and safety of the small town, and hung around with a group of friends from school, doing what young boys in small towns have done since small towns and young boys were introduced. They rode their bicycles, chased cats up trees, got their pant legs tangled in bicycle chains, fell out of trees, caught horned toads, and generally had a good time. Jack, as an adult, admitted that at that time of his life, he only had two short-range goals: passing the next test in school, and having something fun to do on Saturday. He couldn't remember any long-range goals.

For indoor entertainment, Jack had Lincoln Logs in huge quantities and an Erector set to feed his architectural and engineering appetites. Any additional urges to make something were satisfied by an ingenious set of toy equipment that allowed a youngster to melt lead in an electrically heated ladle, then pour the molten metal into molds to make lead soldiers and Indian figures. This simple description of a popular toy of the day is enough to cause a modern reader to dial 911 before leaping out of the chair and running out the door to find the nearest EPA office. If a child was "caught" playing with this toy today, the house would be condemned, torn down and buried in a contaminated landfill, the children checked into a special re-hab center for chronic lead poisoning, and the parents sent to prison for life. In the 1930s, the Kilby's response to questions of toy safety, by Jane's recollection, was "it will help you learn to be careful." Jack didn't seem to show any lingering effects from this outrage.

Great Bend was of a size that it contained a movie theater, and Jack, along with a large percentage of the youthful population, took advantage of it for the special Saturday Matinees. The feature films were followed by "serials," which were short, continuing adventure stories, designed to take you to the edge of your seat at the end, and begging you to come back the next Saturday to see how the hero or heroine escaped from the deadly circumstances. And it worked. Jack confessed to being thoroughly caught up by the serials, especially the "Green Hornet" and "Hopalong Cassidy." As far as feature films, Jack reported that the historical "Clive of India" was far and away the favorite of his youth. Could this have been a youthful case of foretelling of the future? In 1944 he would become "Pvt. Kilby of India."

And, of course, Halloween in a small town brought boys out in droves for some mild yearly mischief-making. One year, Jack and three of his grade-school cronies had a fling at a prank that was destined to turn into family lore. By today's standards, it would rank near zero on a crime scale, but then, in Great Bend, Kansas, four youngsters in the act of dismantling a perfectly good picket fence aroused the suspicions of a neighbor. Halloween or not, the police were called, and in the resulting scramble to escape, one of the perps was caught. The other three, including Jack S. Kilby, escaped cleanly into the darkness.

But mere freedom from the clutches of the police didn't satisfy Jack. The next morning he was seen with the three other young men reassembling the picket fence good as new. Whether this was due to conscience or police action is unclear, but from all later indications of Jack's moral fiber, the scales would tilt heavily towards the conscience side, attempting to repay what could be called a temporary deviation into mob mentality the night before.

Jack joined the Boy Scouts in Great Bend, and was privileged to have participated in a singular event in the scouting world. Jack and others from his troop attended the "1st National Jamboree," held in Washington, D.C., in July, 1937. This huge event was held in the heart of D.C., with the headquarters at the foot of the Washington Monument and camps spread over the center of the city and along the banks of the Potomac River and the Tidal Basin. The Kansas scouts were grouped on Hain's Point, on the Potomac water's edge and, along with 28,000 other scouts and leaders, had a historic ten-day never-to-be-forgotten adventure. The adventure also included a day and a half of heavy rain, which turned the grounds into a quagmire and added immensely to the stories that were brought back to dusty Kansas. Jack, gazing across the Potomac, could never have imagined that he

Jack, back from the Boy Scout Jamboree

would years later be invited to the White House for recognition ceremonies with three U.S. presidents.

Summers have always held special treats for school children, and Jack was fortunate to have been given opportunities to see and do things that would forever affect his life. During the summers he got his first up-close look at what his father did for a living when Jack accompanied him on his inspection tours of the Kansas Power Company's empire in western Kansas. Jack was given an insider's view of the magic of electricity—the power plants that generated the power, the transformer substations, and the hundreds of miles of transmission lines serving the sparsely settled rural areas and small towns.

Jack was also exposed to the negative side of the power business one summer when he was hired to work in a small Great Bend power station that was not associated with his father's company. Jack related this experience with considerable distaste, although it's a good bet that he made no complaints at the time. The work he did was performed by the lowest member of the work totem pole, which is why high school students on summer break were employed. Of the two jobs that Jack mentioned, and there may have been more, the worst was cleaning the tar from the insides of huge oil tanks that were used to hold the heavy fuel oil that fired the boilers in the steam plant. Scraping the sticky black goo from the insides of the tanks, which were exposed to the Kansas summer sun, must have rivaled Dante's descent into Hell. The other, maybe marginally better, job was climbing inside boilers which were out of service for maintenance, and scouring the mineral scale that had built up on the boiler tubes. The lack of black goo was offset by having to work in the boiler room, which didn't need the Kansas sun to be an indoor inferno. Jack, as usual, summed it up briefly but completely, "Everything I did involved heat."

The highlight of summers for Jack and Jane were their visits to the grandparents in Mackinaw, Illinois. Their grandfather, Phillip St. Clair "Dick" Kilby, had died about the time Jack was born, but

Grandma Anna Bell Kilby still lived in the house where Jack's father, Hubert, was born. Her father, Thomas Shaw, who lived to 94, was living with her. Anna Bell was described by Jane as being a tiny woman who wore her hair pulled back tightly from her face. Living as a widow and caring for her father had not been easy and, as Jane recalled, "If she ever had a good time, we never knew it."

Frank and Molly Freitag, Jack and Jane's maternal grandparents, also still lived in Mackinaw. Somewhat the opposite of Grandma Kilby, Grandma Friday, as she was called, was a cheerful sort and liked to have fun. She had her hair "permed," drove a car, and played bridge, all verging on risqué in that place and time. Her kitchen was up-to-date modern, with a bright yellow enameled wood-burning stove side by side with an electric range. But there it stopped—she preferred an icebox to the new electric refrigerator, dismissing this so-called step forward in technology with, "It has to work all day just to make two trays of ice."

When the carnival came to Mackinaw while Jack and Jane were visiting, they were treated as only grandparents can treat grandchildren and get away with it. Grandpa Friday would be the first in line to buy yards of tickets for the rides, (a Ferris wheel and a merry-go-round) and give them to Jack and Jane. They would ride to their heart's content and beyond, and then sneak the remaining tickets to their friends. A hard day at the carnival might end with fondly remembered ice cream sodas personally made by Grandpa. It was good to be a grandchild in Mackinaw.

Chapter Three

High school

Jack grows tall, tries everything in high school, and finds a long-term interest.

MAY 20, 1937, WAS A GOOD DAY. Jack successfully completed junior high school in Great Bend, Kansas. He was prepared for high school and the opportunity to broaden his activities, both within the classrooms and without. Jack may have been verbally frugal in high school as he was as an adult but, if he was, it didn't keep him from participating with enthusiasm in a variety of activities outside of the classroom.

The high school classrooms also kept him busy, with a final tally of two years of Latin, three years of science (biology, physics, and chemistry), American and state history, geography, four years of English, and one and a half years each of algebra and geometry. Jack took no trigonometry because it wasn't taught, and this was to cause Jack's college plans to take a major detour.

Jack had another talent that had begun to show up a few years before—Jack knew to how grow. He had outstripped his classmates early on, and by high school he was near his adult height of 6 feet 6 inches. He had the physical size to interest any high school sports coach, and the Great Bend High coach responded right on cue. After playing football his sophomore year, Jack switched over to basketball during his junior year. Jack later commented that he couldn't see much use in running back and forth in a small area, although Jane suggested that if it had been explained to Jack in mathematical terms, he might have found it interesting. But the surprising aspect of these extracurricular activities was that Jack tried things and did things that now seem incredible to those who only knew Jack Kilby as an engineer. Few would believe that Kilby played either football or basketball in high school, much less both. Maybe curiosity isn't the right reason, but Jack expanded his young horizons and got the most out of his teenage high school experiences. He showed a remarkable boldness as he was growing up.

During this time of poking about at life's experiences, Jack hit upon a keen interest that he maintained for most of his life—photography. The exact beginnings of this interest that bordered on a passion at times are unknown, although his father was an amateur photographer, which gave Jack an early exposure to the art. The family also had a 16mm home movie camera, rare at the time, and captured the early antics of Jack and Jane.

Jack expanded his photographic realm in high school as he built a home-made darkroom, and gradually equipped it with the beginning tools of the photographer's trade. Soon he could develop his black and white films, print them, and make enlargements. Beginning as a sophomore, he was in the Great Bend High School Camera Club, and served as its president his senior year. In his senior year he was also on the staff of the yearbook, the *Rhorea*, as photographer. Jack advanced quickly from beginner to competent photographer, and he was going to get a lot better.

Chapter Four

The Great Blizzard of 1938

The unexpected result of bad weather in Kansas

SOMETIMES A NATURAL OCCURRENCE can become a life-changing event, and the devastating Kansas blizzard that occurred April 7 and 8, 1938, was one of those. Jack Kilby had not, at that time, settled in his mind where his life was going. Years later his sister Jane said she thought Jack was leaning towards photography as a career, based on his great interest and activity in it, and his early success. Jack was a very good photographer for a wet-behind-the-ears teenager, and by all indications it was becoming his passion in life. And then along came the blizzard.

This late spring blizzard hit hard and caused major disruptions in the western half of Kansas, including outages in the electric service of the Kansas Power Company. Hubert Kilby immediately began the struggle to assess the damage to the power system that he was responsible for, and to dispatch repair crews. This struggle was futile, since the telephone lines were also down, and communication in and out of Great Bend was virtually cut off.

Hubert, with son Jack in tow, trudged the few blocks through the snow to his friend, J. R. "Roy" Evans, who he knew was a "ham," an amateur radio operator. Roy was already preparing to swing into action to help with emergency communications as hams had done in the past. At that moment, hams all over Kansas were gearing up to do the same thing. Although little known and little appreciated, amateur radio operators around the world were, and still are, prepared to act as a last line of communication in the event of regional disasters.

Roy first contacted another Great Bend ham, Richard Livingston, and then they both began making contacts with other amateur radio operators in the seriously affected areas of western Kansas. Hubert worked through Roy's station, W9DKI, and was able to set up communication schedules with his power plant managers, and to check the condition of the Kansas Power Company's power plants and power lines in Concordia, Jetmore, and other beleaguered towns. Before the mission was over, four amateur stations in Great Bend had been on the air night and day, trading emergency messages with other ham stations in Kansas. An operator in the town of Larned sent the report that drifts were eight feet deep, and all attempts to clear the roads by the Highway Department crews had failed because of the blowing snow.

Jack said years later in a conversation with Roy Evans' grandson, physicist Dr. Donald L. Walters, that this first contact with amateur radio was a major influence in his life, and "certainly the beginning of my interest in electronics. It convinced me that I wanted to study electrical engineering."

The blizzard brought a new and abiding interest into Jack's life, and he responded by gathering information and parts and beginning the construction of his own ham radio station. He bought books from the American Radio Relay League, the amateur radio national organization, learned Morse code, and studied the fundamentals of electricity and radios. Then he was driven to Denver by his father to take the Amateur License exam—pages of technical questions and a sending and receiving Morse code test at 13 words per minute. After a long two months, Jack's amateur radio station license arrived in the mail. He had passed and was now W9GTY.

Jack was an active ham during his junior and senior years in high school, but was only occasionally active on the short wave bands after he went off to college. Ham radio had accomplished a great deal in the Kilby family—it helped get electrical power to snowbound Kansans and started a young man on his path to change the world.

There is a postscript to this story. On September 30, 2002, Dr. Donald Walters, the grandson of Roy Evans, arranged for a reprise of the blizzard events. The FCC reactivated Jack Kilby's call, W9GTY, and, using borrowed equipment, Jack was able to make ham radio contact with Donald some 64 years after the event. Donald was using the original National FB-7 receiver that his grandfather had used in the blizzard of 1938.

Chapter Five

High school ends

Jack does well, graduates from high school, and then runs away from home.

Jack Kilby's 1941 graduating class at Great Bend High School included about 140 eager students, and none of them could dream of what the 7th of December of that year was going to bring. But graduation did bring them their *Rhorea* yearbooks, with the excitement of finding their photos and just dying to get that cute boy or girl to sign it with some juicy tidbit. No one could have dreamt that the 1941 *Rhorea* would later prove a goldmine to anyone delving into the archives for information about young Mr. Kilby.

Jack, the human bean-pole, is in two group photographs of the *Rhorea.* The Student Council page has the nineteen members in two long gender-sorted rows backed by a brick wall. The photographer's dilemma could be imagined as he tried to fit the out-sized Kilby into the disparate line of teenaged boys. His "aha!" moment was realized when he placed Jack in the center of the back row, and fitted ever-shorter boys to his left and right. The effect is something like a peaked-roof house. The fact that Jack was on the Student Council tells two things about him that might otherwise be missed: Jack was a pretty good student, and he had the respect of his teachers and classmates. They knew who he was and liked him.

In the second photo are the members of the National Honor Society, whose text advises us that pupils elected thereto must be in the upper third of the class, and not more than 15% of any senior class is eligible. Again, the photographer appeared to be puzzled by the Kilby stature, and this time placed Jack on the far right of the photo with a somewhat symmetrically arranged tall student on the opposite end. The back row now resembled an overloaded line of washing, with Jack as one end post. Even with this selection of sizes, Jack is a full head taller than the boy next to him. Size notwithstanding, Jack fit very well into their requirements of scholarship, leadership, character, and service. Might be Nobel material some day? From Great Bend High School? Really? Nah.

In the listing of "Senior Records" in the rear of the yearbook, Jack Kilby was somewhat surprisingly identified as a freshman and sophomore member of the Red and Black Boys, a "spirit" group for males, and the counterpoint to the female Pepperettes. He was in the band, playing trombone, from freshman through junior years, and he entered unidentified "Music Contests" his junior and senior years. He is listed as participating in the Operetta during his senior year, but the type of participation is not identified. Somehow the picture of Jack singing in an operetta doesn't easily come to mind, but who knows? Jack brought several other surprises out of his high school years.

Jack's photographic activities in the Camera Club and *Rhorea* staff were noted, but his membership in the National Honor Society inadvertently omitted from the Senior Records list.

His election to the Student Council is identified, as is the fact that he was a member of the Senior Class Play, working appropriately enough as the electrician. This leaves his participation in the Allah Rah, which Jack dealt with as a senior, and whose charter and purpose shall ever remain a mystery.

Jack had an active high school life, and he brought it to a close by graduating in the spring of 1941, to the cheers, we can be sure, of Hubert, Vina, and Jane. He did good.

Then, like many graduating seniors, Jack and two of his friends sprang free to celebrate the brief period between the ending drudgery of high school and the beginning drudgery of either college or work

with a big adventure. The left photo shows Jack standing by the "hot-rodded" 1930 Model A Ford roadster used in this bold exploit. A closer examination of the vehicle shows creative "improvements," such as the removal of useless fenders, installation of what might be the embryonic outline of a pickup truck bed in the back, and "Supercharged" painted on the hood.

In the summer of 1941, the three daring youths first drove this engineering marvel to Florida, which was the destination of one of the young men, and probably the lame excuse to make the trip in the first place. From there, Jack and his remaining friend drove up the coast to Washington, D.C., stayed briefly, and returned to Kansas. It was reported that the last leg of the return trip was done with little or no operational brake linings on the car, and stopping had become an exercise in down-shifting and foot-dragging.

Something over 60 years and a lot of technology elapsed between these two photos of Jack St. Clair Kilby. The recent photo of Jack was taken in the fall of 2004 on the occasion of his 81st birthday. His two daughters, knowing of his youthful adventure in his souped-up jalopy, found the means to give their father a bit of a replay of that heroic trip, courtesy of Don Grubb of the Dallas Model A Ford Club.

Chapter Six

College bound

MIT, here we come! Well, maybe not. Jack takes his second choice in colleges.

WELL, NOW WHAT? Both of Jack's parents had graduated from the University of Illinois, but Jack had no particular attraction to the school. He had decided to pursue a degree in electrical engineering, and had picked the Massachusetts Institute of Technology, MIT, as his first and only choice. The story of his failure to gain entry into this top-notch school has been told and retold. Some of the facts of these events are muddled by time, but it seems that MIT required college board tests for admission, and Jack took them early in the summer of 1941. He failed the boards, being deficient in the mathematics portion, undoubtedly related to his lack of high school trigonometry. Later in the summer, he went by train to MIT and first took a remedial math class that they offered, and then retook the boards. He scored 497, again short of the required minimum of 500.

By the time Jack arrived back in Kansas, it was only a week before the fall semester would begin in most colleges, and not enough time to make an application by mail. Hubert piled Jack in the car, and they drove to see what the University of Illinois could do on short notice. Maybe it was the power of having not one but two alum parents, or maybe it was Jack's transcript and extracurricular activities in high school, but in any case Jack was signed up to begin his classes in the electrical engineering school. Jack later admitted he never actually applied for entrance to the University of Illinois—he just enrolled and began going to classes. Whatever works.

Chapter Seven

University of Illinois: Part I

Jack settles into college class work and then gets unsettled on December 7, 1941

JACK KILBY BEGAN WORK on his degree plan to earn a Bachelor of Science in Electrical Engineering at the University of Illinois, Urbana, in the fall semester of 1941, and pretty much vanished from the history radar for two years. He joined the Acacia Fraternity, and lived in the frat house. Later he would become a life member of Acacia. The frat house had no provisions for his amateur radio equipment, so Jack joined the Ham Club on the campus, and occasionally used their station to "work" a few hams around the world.

There is information from Jack that his first two years of college had little relating to engineering, electrical or otherwise. His later Military Service Record verifies that information, stating that Jack's civilian occupation was:

> *"STUDENT, COLLEGE, ELECTRICAL ENGINEERING —Completed 2 years of college at University of Illinois, Vrbana, [sic] Illinois, majoring in electrical engineering. Studied such subjects as mechanical drawing, algebra, trigonometry, calculus, chemistry, physics and wiring."*

Jack finally got his trig and apparently "wiring," but the good stuff would start his third year. He managed to feed his photography habit, as the 1943 *Illio*, his sophomore yearbook, lists Jack on the *Illio* staff as an assistant photographer. Pretty good for kid from a small town in Kansas.

One thing stuck in Jack's mind, and well it should have. He was in the Acacia House when he heard the news of Pearl Harbor, the beginning of World War II. He had not even completed the first semester of his college career. This must have shaken even the unflappable Jack. It didn't take a fortune teller to predict that his four-year path through the University of Illinois was going to take a serious detour.

On December 12, 1942, Jack signed up in the U.S. Army Enlisted Reserve Corps. He was now well into his second year of college, and this may have been done with the agreement of letting him finish it. Jack completed his sophomore year and entered active duty in the U.S. Army on June 12, 1943. He was 19 years old.

Chapter Eight

Kilby goes to war

Kilby joins the Signal Corps and then the OSS. He travels to India, Burma, and China, and then comes home. Thank God.

JACK WAS TRANSFERRED to active service, and was ordered to report to Ft. Leavenworth, Kansas, on June 12, 1943. This left only a brief time in Great Bend to, once again, say his goodbyes to his parents after completing his sophomore year at the University of Illinois. Only this time, the goodbyes were for their only son going off to an unknown fate in a world at war. What a difference from the goodbyes of two years earlier.

Jack's college degree plan in electrical engineering led to his being assigned to the U.S. Army Signal Corps and basic training at Camp Crowder. World War II found Camp Crowder, near Neosho, Missouri, the destination of most of the Signal Corps recruits.

Camp Crowder was originally built in 1941 to be an infantry division training center but for one overlooked detail. As it was being completed, a civilian engineer determined that a large Shell Oil Company pipeline cut through the middle of the proposed artillery impact area. The soldiers and their cannon from the Second Army that had already arrived were sent back to Louisiana, and the base turned over to the Signal Corps, much to the relief of the pipeline.

The camp was immense, sprawling over nearly 43,000 acres of Ozark foothills. It was a city of soldiers. The Midwest Signal Corps

School alone could accommodate 6,000 students, and Jack Kilby was now one of them.

After completing three weeks of basic training, Jack began his classes in radio operation. His discharge papers show that he attended and completed "Low Speed Radio Operator School." Jack, for reasons not disclosed, did not like Camp Crowder nor the Signal Corps training program. And how much did he dislike it? Enough to try an extraordinary way to get out of it.

The Office of Strategic Services, better known as the OSS, was actively recruiting the Signal Corps trainees at Camp Crowder. The OSS was new as of World War II, and would later be renamed the CIA, the Central Intelligence Agency. A contemporary account related that the OSS made announcements over the Camp Crowder mess hall loudspeakers that encouraged qualified radio operators to volunteer for immediate overseas assignment. Many considered but few made applications, because there were several catches to the qualified part—two years of college preferred, some training in a European language, first choice French, and the desire and skills to become a parachutist. Combine all this with the likelihood of a dangerous assignment, and it made a big lump to swallow. But if you could choke down the requirements, you could come in for an interview, and Jack did.

During his interview, Jack was told by the OSS they were putting together two-man teams to parachute into occupied France. Jack's story, in his own words, "I knew I wasn't too well-designed to just blend into the landscape, but if they had anything else, I'd be glad to do it." How's that for a creative way to escape Camp Crowder? Not only succinct but polite. Probably propelled by his two years of college pointing towards an electrical engineering degree and ham radio experience, plus his Camp Crowder radio operator training, they took him up on his offer. Besides, how could you resist a 200-pound, six-and-a-half-foot-tall volunteer who already knew Morse code? Jack became a Private in the OSS with the promise and expectation of repairing their radio gear, wherever in the world it might be.

After more specialized training, Jack was assigned to Detachment 101 of the Office of Strategic Services. The OSS had been formed in April, 1942, for "covert operations, such as unconventional warfare behind enemy lines." They were the precursors of the now familiar "Green Berets." Detachment 101 of the OSS was in the China, Burma, and India, or CBI Theater to neutralize the successful Japanese jungle tactics.

OSS Det 101, as it was referred to in typed orders, organized and put guerilla warfare into action on the Allied side, and pioneered using radios for communications and coordination with friendly rebel troops. It was a small group of specially trained U.S. soldiers, about 1,000 at its peak including 250 officers. But these few commanded a monstrous invisible army of skilled and well-equipped local guerillas who numbered almost 11,000 men. By the end of 1943 the OSS had six intelligence-gathering bases in northern Burma, all communicating by radio with their main base in Nazira, India. This was the strategic cluster of OSS Det 101 forces preparing to begin the first of three major campaigns they were to wage—to oust the Japanese Air Force from their field near Myitkyina in northern Burma and then capture Myitkyina. This Japanese airbase was responsible for regular fighter attacks on Allied cargo planes flying the "Hump" and bringing desperately needed supplies into China. It was the top priority of OSS Det 101 to clear the Japanese flyers from this area, and that's where Jack was going. Radios were the life-lines in the jungle-like terrain, and his job would be to keep them working.

The "water transport" for Jack's overseas duty left the U.S. dock on March 9, 1944, headed for India. He arrived on the 10th of April and reported to his duty station at Chabua, India, near the northern boundary of Burma. This outpost was less than 50 miles from the OSS radio communication center in Nazira.

Reports were that fighting was desperate, but Merrill's Marauders, one of the elite OSS-led guerilla forces, seized control of the Japanese-held airport near Myitkyina on May 17. Capture of the town

Kilby in the doorway of the "Brooklyn Edison," where the base generators were located.

of Myitkyina would prove harder, and the OSS "Galahad" thrust took the city on the 3rd day of August, 1944, after two months of bloody fighting. The first CBI OSS major campaign had been accomplished.

Shortly afterwards, Private Kilby accomplished his first advance of the campaign also. He was promoted to Tech 5, equivalent to an Army Corporal. His work at the Chabua base was good.

A month and a half later, in the middle of October, 1944, T-5 Kilby was sent into the recently liberated Myitkyina for a temporary duty assignment. He was sent with explicit instructions "upon completion, return to proper station." There is no information to identify the reason for Jack's sudden out of town "service call," but as the Allies were scrambling to put Myitkyina back together after a two-month siege, it might have been useful to have an electrical and radio guy handy. And Jack was a handy guy.

The war for Burma raged on and, two months later on 15 December, the Japanese evacuated Bhamo, a key stronghold between the Allies and the crucial Burma Road. Within a month, OSS patrols were sparring with Japanese troops on the Burma Road itself. The OSS and the advancing Allied troops kept the pressure on the tiring Japanese army, and on the 7th of March, 1945, the Japanese troops pulled out of Lashio, Burma. This removed the last major obstacle and opened the way to the recapture of the Burma Road. The second OSS campaign was completed. Now for the third—open the Burma

Road for Allied use to move supplies eastward and establish bases deep in China.

How did Jack get a Texas Instruments jeep?

The evacuation of Lashio allowed the Ledo Road, built by the British a few years before World War II, to link up with the Burma Road, and OSS troops were now able to set up new bases in China. The third and final OSS Detachment 101 campaign had been achieved, and its role in India and Burma was winding down as the fighting moved east into China as the Japanese retreated.

Jack's radio repair duties were also winding down. He received orders stating that his responsibilities to Det 101 were ending as of 5 May 1945, and that he would be transferred to OSS headquarters in Kunming, China. He was to report "on or about 20 May 1945." What they didn't tell him was how he was going to get there. Besides being a top-notch electronic technician and radio operator, Jack had a skill that was in short supply—he could drive a truck.

There was now a steady stream of troops and materiel moving by truck over the recently liberated Burma Road. The fighting had moved to China, and the U.S. Army was going after the enemy with the intention of pushing him off the continent and into the sea. Cargo aircraft were still flying the Hump, but the bulk of the supplies and men took the low road, the Burma Road, to Kunming.

Jack did more than hitch a ride to Kunming. Jack drove a 2½ ton truck, better known in the Army as a "deuce and a half," or "6x6," to Kunming. Other WWII reports of driving on the Burma Road are mixtures of episodes of terror alternating with stories of nightmarish quagmires, of switchbacks so severe that the trucks had to be see-sawed back and forth to get around the turns, and bottomless drop-offs on the narrow rutted road. Something over 700 miles of this precarious track led from Lashio, Burma, to Kunming, China, and every mile both a triumph and a misery.

Jack's stay in Kunming was brief, as orders were promptly cut to move Jack far inland to the city of Hsian, known also as Sian, and now known as Xi'an. Near the middle of China, the OSS had recently established a major communication base at this central location, and Jack was to be stationed there. Aircraft shuttling supplies and men between Kunming and Hsian were in short supply and loaded to the gills, so for transportation Jack was assigned a seat in a jeep in an OSS supply convoy leaving for the new communications base. This trip from Kunming to Hsian would make the Burma Road seem like a short drive, as it was nearly 1,500 miles of poor road, most of it mountainous, and much of it spotted with armed bandits. The retreating Japanese had been replaced by free-lance thugs, who were either renegade Chinese troops or opportunistic civilians.

Jack was to be the passenger and radio operator in the convoy's lead jeep, driven by Oliver Caldwell, an OSS officer. Caldwell later wrote the book, *The Secret War: Americans in China 1944–1945,* and described the experiences of this convoy to Hsian. Caldwell's book contains two paragraphs that are of special interest:

"About forty miles out of Kunming, we felt that we had gotten rid of the bugs and were rolling along as a unit. We had fallen into a routine. As we skirted the side of a valley, I noticed that my radio operator, Jack Kilby, the other passenger in my jeep, had fallen asleep with his head bent over his knees. I was sleepy myself.

Suddenly a bullet whipped past the back of my neck. I had the impression it might have gotten Jack if he had been sitting up straight. It was hard to believe that anyone was actually shooting at us, but my momentary doubt was removed when another bullet ricocheted off the road in front of me. I could either stop the convoy, order the men into the ditches, and fight it out, or try to run the gauntlet. The road was good so I chose to run for it since the firing was coming from trees several hundred yards above us. The fact that I had no brakes didn't give me much choice. We were lucky and suffered no losses."

Saved by a nap? Who knows. The 12-day caravan described in detail by Caldwell was long, arduous, and dangerous. Jack arrived for duty at the Hsian OSS compound on the outskirts of the city on June 9 to begin his new Army job as Radio Operator. The OSS had an active and continuing interest in the interception of enemy radio transmissions and was still parachuting agents into hostile territory. Kilby, as an OSS radio operator, could have been tapping into the Japanese army's battlefield communications or talking to a fellow soldier behind enemy lines.

What is known is that Kilby did an outstanding job. On the 24th of August, 1945, just after World War II had ground to a bloody end, Jack received a promotion to T-4 Radio Operator, a scant two and a half months after beginning his new job at the Hsian base. The promotion read that Jack was "excellent" in both character and efficiency, and that this promotion was "highly recommended by the station." Jack was now the equivalent of a buck Sergeant in the Army.

That welcome piece of military paper was followed a month later, September 30, by an even more welcome one. Jack received orders relieving him of his duties, and commanding him to report to the

Replacement Depot #3, Camp Angus, Calcutta, India, on or about 5 October 1945, "on the first available transport from APO 627," the Kunming OSS base. From the "Repl Depot" he was to catch the first available water transport to Seattle, Washington, in the good old U.S. of A. He and a whole lot of other GIs were through with military service and were going home.

The *Marine Raven* sailed down the Hooghly out of Calcutta on the 12th of November, 1945, with Jack among the 2,627 passengers. The 520-foot troop carrier headed for Seattle and home for the human cargo and crew. It was reported in the shipboard newsletter, *The Raven Times*, (Saturday, 8 December 1945, Souvenir Edition) that the passengers had all been properly accommodated by using "sardine-packer's efficiency." News that they didn't already know was that they were scheduled to "raise Cape Flattery Tuesday afternoon, and after an 8-hour sail on Puget Sound, dock in Seattle." It would be a 9,500 mile journey of 28 sailing days and 34 days aboard, according to the newsletter. This time and distance included dodging a serious storm in the Pacific, and changing their destination from Seattle to San Francisco and back again,

The end of Jack's military service came as an anticlimax. He arrived at the Separation Center, Ft. George G. Meade, Maryland, on December 19, and was successfully separated two days later. Jack St. Clair Kilby was honorably discharged from the services of the Army, "at the convenience of the Govt," on Christmas Eve, 1945. It had been a long and difficult interruption.

This chapter tracing Jack's tracks through China, Burma, and India was put together with information in his files that are now in the SMU Kilby Archives. Jack had saved most, but not all, of his military orders and promotions. Overlaying Jack's time line with the time line from the internet on-line history of the OSS Detachment 101 made a mostly verified and plausible story, even with a few missing links. And then the excellent "Biographical Memoirs" of Jack Kilby written by one of his best friends, Charles Phipps, came to light. In it, Kilby had told Charles a war story that on a long convoy they had to stop within 40 or 50 miles of Mao's Red Chinese army camp and wait for orders from Washington as to what to do. They waited for "a couple of weeks," and then were told to turn back and not confront the Chinese army. Was this the convoy from Kunming to Hsian? Rereading the history of the Detachment 101 and Caldwell's book brought no enlightenment. Just when you think you've got the story straight…

Kilby, left, dines in the elegant "Bamboo Palace," Chabua, India

Chapter Nine

University of Illinois: Part II

Back to college, full time and then some, Jack earns his BSEE degree and a job at Centralab

HOME FROM THE ARMY, Jack wasted no time in getting back to the University of Illinois and picking up his education where he had left off in the spring of 1943 after completing the first two years of his degree plan. As before, Jack vanished into the maelstrom of the University of Illinois, only this time with boat-loads of veterans. He began his junior year and wasn't heard from until he graduated August 5, 1947. One reason for his lack of communications may have been that he restarted college in the spring of 1946, and went to college essentially full time from then, summers and all, until he graduated. He did not take any part-time work because of his course load, and because, he said, "I was on the GI Bill and it was almost enough money." Jack completed the last two years of his Bachelor of Science in Electrical Engineering in a year and a half.

He did mention that he'd done a little "hamming" at the amateur radio club during the year and a half, and he jumped back into his beloved photography and was on the *Illio* yearbook staff again as a photographer.

Jack's story came back into focus as it neared the time to finish college and go to work. Getting a job suddenly became a priority. From information he had gleaned from his military and ham radio experience with electronic equipment, he sorted out 30 or so companies, and sent them job inquiry letters. Jack had only three responses requesting on-campus interviews—Collins Radio, Centralab Division of Globe-Union, and a group from Bell Labs. He also did a walk-in campus interview with General Electric. Jack later commented on his lousy batting average with his inquiry letters. What Jack may have forgotten was that few people, excepting a handful of top scholastic performers, received any more interest from electronic companies than he did. The first years following World War II were not the best for electrical engineering graduates.

Collins Radio was a well-known and prestigious firm, and Jack went to Cedar Rapids, Iowa, for an interview. They expressed interest in Jack, but didn't make him a job offer. The Bell Labs group had nothing that seemed attractive to Jack, and he did not pursue their response to his inquiry. In the meantime, Jack continued talks with Centralab in Milwaukee. They wanted Jack to join the group that was making silk-screened circuits to reduce the size and cost of circuit component clusters. This somewhat off-beat bit of design and engineering appealed to Jack as the most interesting field he had encountered, and after graduation he accepted the offer of employment by Centralab Division of Globe-Union, Inc., in Milwaukee, Wisconsin.

Centralab had some familiarity with electronic miniaturization, and because of this, the government had called upon them near the end of WWII. The government was desperately trying to ramp up production of the super-secret proximity fuses. These fuses were developed late in the war by the U.S. in cooperation with the British, and were a giant technical leap. Fitted in the nose of a cannon shell, these fuses used a kind of radar to detect the proximity of an object and cause the shell to explode even though it had not struck the target. That is, if the fragile vacuum tubes and electronic components had not been scrambled like an omelet by the starting shock of firing the cannon. It had been an incredibly difficult engineering and manufacturing project, but the results were extraordinary.

The fuses greatly increased the effectiveness of the radar-controlled antiaircraft batteries on the British coast, and were the edge needed to finally stymie the German V-1 "buzz-bombs" coming across the channel to detonate in England. In the crucial Battle of the Bulge, ground-based Allied artillery used the proximity fuses to accurately control lethal shell bursts low over German divisions in the open. It was estimated that they increased the effectiveness of the artillery by seven times. But the most urgent use of the proximity fuses late in the war was in the Pacific Theater, where the fuse was the only weapon that was effective in combating the continuing decimation of the U.S. fleet by the Japanese kamikaze air attacks.

In an attempt to increase the production of proximity fuses, the National Bureau of Standards contacted several companies seeking novel ideas for building these miniature electronic circuits. Centralab got the contract in 1945 and began work and, when the war ended, the NBS continued funding for the project. This was the company and the group that hired Jack in 1947. Jack and "miniaturization" met for the first time.

Chapter Ten

Jack miniaturizes things

Kilby learns to make things smaller while expanding his education with an MSEE

JACK'S ARRIVAL AT CENTRALAB marked a timely entrance into the field of electronic miniaturization. The manufacture of proximity fuses was only one of the many problems that had surfaced during World War II.

The days of the vacuum tube had come to the point that the reach of electronic desires was exceeding the grasp of existing hardware. The military in particular had problems with airborne electronic systems. These fighting systems, which had grown exponentially during the war, had swelled to hundreds of vacuum tubes and threatened to turn the new military airplanes into the aluminum equivalents of the dodo bird.

And it wasn't just the size and weight that was reaching a dead end. The reliability of systems with thousands of parts and thousands of electrical connections had been dropping like a rock. The famous "tyranny of numbers" phrase describes it well—the desire to build ever more complex (and ever more useful) electronic gadgets was running out of steam—it was collapsing under its own weight. The military were the first to suffer, as the B-29 Super Fortress had systems with almost a thousand vacuum tubes and untold numbers of electrical parts. The military were leading the way in finding a solution to this quagmire of complexity. The need was to shrink the size and minimize the number of connections, and break down the "tyranny" barrier to building complicated but reliable electronic systems.

Jack's first work at Centralab was centered on "silk screen circuits." This miniaturization technique had already been developed and brought to a successful production level by Centralab. It replaced the individual wires connecting electrical parts with stripes of conducting paint made with silver powder. This "painted circuit" process was done with a variation of the "silk screen" printing method, commonly used at the time, in which silver "ink" was squeegeed through a pattern on a porous screen onto an insulating surface, generally a ceramic plate. Originally, silk cloth was actually used because of the toughness and fine weave. This process is where the term "printed circuit" originated, even though the circuits are no longer printed. Centralab was a pioneer in the early techniques that grew into the modern "PCB," the Printed Circuit Board, that is the basic building block for the vast majority of modern electronic equipment and systems.

Centralab had developed a successful line of specialized products for the fledgling electronics industry to simplify the construction of vacuum-tube circuits. A manufacturer of a radio or TV, for example, could eliminate handling and wiring a handful of parts by purchasing a specially made module from Centralab and wiring in one part instead of many. Kilby's work in this expanding area of engineering over the next ten years included a key side trip to meet Ma Bell.

U.S. Patent 2,637,777 was filed on February 27, 1950, and issued May 5, 1953. The inventors were Jack S. Kilby and Alfred S. Khouri, Jack's boss. It was Jack Kilby's first patent, but far from his last. The title was "Network Having Distributed Capacitance," and was a unique idea for making a very compact and reliable electrical filter using the existing Centralab screening techniques for miniaturization. Jack's inventive brain was off and running.

Jack was also off and running in another direction shortly after he began work at Centralab. He began night classes for a master's degree in electrical engineering from the University of Wisconsin's Milwaukee Extension Division. He described this adventure as "kind of a mail-order master's degree." A grade sheet from the University of Wisconsin dated "2nd Semester 1950" shows Jack registered as a 5th year student, receiving credits for two classes, EE200 and EE215, and scoring A's in both.

He received his Master of Science in Electrical Engineering in 1950 from the University of Wisconsin, Madison. His master's thesis was "Design of a Printed Circuit for a Television IF Amplifier." Right down his line of work. In 1952, Centralab was selling a "TV IF strip molded unit with tube socket and all RLC components to fit a PC board." It had the advantages of "minimum connections and parts, low labor, quick service, and it looks better." Form and function nicely combined.

The consumer product work at Centralab was growing as the post-war economy gained traction, and their products included vacuum-tube hearing aids in addition to the miniature component packages. They had the hang of making things small, and things were soon going to get even smaller.

In early 1952, Jack was told to pack his suitcase and prepare to spend a week and a half in New Jersey. He and Bob Wolff, his boss's boss, were going to a symposium on transistors at Western Electric's Bell Laboratories. Centralab was going to learn how to make those things for their hearing aids.

Chapter Eleven

Bell Labs and transistors

Bell Labs teaches Jack how to build transistors. Jack gets Centralab in the transistor business. Jack leaves.

THE TRANSISTOR WAS INVENTED in late 1947 at Western Electric's Bell Laboratories in Murray Hill, New Jersey. When it became apparent that the development of this exciting little device would be faster if the technology were spread around (plus a little nudge from the Justice Department's antitrust lawyers), Bell invited several hundred ("seven bus-loads") of scientists and engineers to attend a five-day get-together the week of September 17, 1951, and catch up on this new and promising technology.

The presentation consisted only of technical papers, and each attendee was furnished with a Bell Labs book, *The Transistor – Selected Reference Material on Characteristics & Applications*, roughly the size of a volume of the *Encyclopedia Britannica.* The 1951 symposium was a smash hit technical event, but it included no practical information on how to build this mysterious new beast—it just whetted the appetites of the scientists and engineers who could now see just how big this thing was going to be. For such a tiny thing, it looked like it was going to be really big.

Bell Lab's next step to fan out transistor technology was to set up a second symposium to transfer the process information, right down to the Do-It-Yourself level. This would be the star event for those companies who were willing to purchase a $25,000 license to get in the transistor business. As an added incentive, the $25,000 could be redeemed as royalty payments if they sold any transistors later.

This How-To symposium was set up for April, 1952, and of the 35 companies who paid for the license, about one third were foreign. Each licensee could send four persons to the symposium. Later estimates indicated about one hundred technical people attended the eight-day affair.

Globe-Union, the parent company of Centralab, signed up for the license and sent Jack Kilby and R. L. Wolff to learn the secrets of turning germanium semiconductor material into transistors. They were promised that this symposium would "enable qualified engineers to set up equipment, procedures, and methods for the manufacture of these products." These products were two: the point-contact transistor and the grown-junction transistor.

The licensee list held a variety of big-name electronics companies and a sprinkling of middle-sized firms with a special interest in tiny amplifiers, such as Globe-Union, maker of hearing aids. There were several small non-descript firms that most people had never heard of—Texas Instruments, for example.

It's not hard to imagine the lanky Kilby squeezing down the row of seats, stepping on TIer Pat Haggerty's toe, and excusing himself, although Kilby later said he had no recollection of the personnel from Texas Instruments. The TI people probably couldn't help but notice the tall guy sticking up above the crowd, but would have had no idea that he would interact strongly with their lives in a few years.

The lectures at the Murray Hill plant were interrupted by a bus ride to Allentown, Pennsylvania, to tour the Western Electric transistor production facility. The eight days of high-pressure technology transfer were summed up by Mark Shepherd, one of TI's attendees, "They worked the dickens out of us. They did a very good job; it was very open and really very helpful."

Back in his small engineering group at Centralab, Jack took over the design and building of the necessary special equipment: a reduction furnace to make germanium, the primary material for transistors, from germanium oxide; a zone refiner to purify it to exceedingly high quality; and, most important, a crystal puller to grow the germanium crystals—the silvery stuff the transistors were cut from. He was the original Jack Kilby of All Trades. In addition to the three big pieces of process equipment, there were dozens of other supporting items that had to be bought or built—dry boxes, microscope stations, electrical testing equipment, ultrasonic saws, and a lot more.

This was to be a slow haul. Jack was still involved in the job he had before he visited Bell Labs and became a transistor expert. He was not able to begin work in earnest on the transistor project until the fall of 1952, and then it took over a year to get the facility set up and running. In his annual engineering report to Wolff at the end of 1954, he reported "a few samples of junction transistors were delivered the first of the year." He expected to be able to produce 2,000 transistors per day in May, 1955, even though "all germanium processing [for production] is done in the engineering lab." By comparison, General Electric produced three million transistors in 1954, and was planning on 250 million in 1960. Texas Instruments beat almost all of them to the market place with the sale of 10 point-contact transistors to the Bulova Watch Company on the 30th of December, 1952.

Jack's semiconductor lab had correctly ignored the point-contact transistor that was rapidly becoming obsolete even as they were sitting in the audience at Bell. The choice to concentrate on the junction transistor was made early. Centralab, under Jack's guidance, produced a handful of experimental grown-junction transistors, but quickly settled on the alloy-junction type for their amplifiers and hearing aids because of the relatively simpler process.

Transistor applications were growing slowly. The largest consumer use of transistors in mid-1954 was in hearing aids. Since Centralab had been an established supplier of the vacuum tube models to the industry, it wasted no time in converting to their new in-house transistors, and Kilby was pulled in to assist in the circuit design also. At the peak of production, Centralab was cranking out 250 hearing aids per week, using a thousand of their home-grown transistors.

With more of Kilby's expert engineering, other transistor products followed the hearing aids. Soon Centralab had several lines of miniature audio amplifiers on the market. A letter from an engineer at Minneapolis-Honeywell to Jack said, "Samples of the highly miniaturized equipment that you showed us here Friday made quite a sensation." It was a four-stage amplifier in a tiny round can little bigger than a pinky fingernail. Small wonder it was a sensation to engineers who were used to seeing vacuum tube amplifiers a hundred times larger.

Jack was doing what he did best—solving problems by inventing, and his inventing frequently involved making things smaller. His invention disclosures on several types of transistor improvements and circuit uses were submitted to the Globe-Union patent attorneys, where they were written up and filed for patents. Two examples are "Fused Junction Transistor Assemblies," filed in March, 1955, and later, "Modular Electrical Units and Assemblies Thereof." Jack would eventually collect eleven patents while working at Centralab.

One of Jack's patents is worth a closer look. It is U.S. Patent #2,823,262, "Telephone Answering Device." This patent was filed October 21, 1953, and issued February 11, 1958, and assigned to H. A. Milhaupt, Inc. This was long before you even heard of a telephone answering machine, much less could buy one, and the drawing in the patent shows it connected to a dial-type desk telephone. It looks like it belongs to the Flintstones, but it was an answering machine. Did Kilby invent the first answering machine? Maybe or maybe not, but he was certainly a pioneer in the field. Centralab was aware that Jack was working on this idea, and had undoubtedly allowed him to keep the rights to it. This break-out from his Centralab work-a-day world was an indication that the inventor wanted to invent. Kilby had a lot of ideas, not just at Centralab, but throughout his life.

In March, 1958, Jack received a letter from the president of Globe-Union congratulating him on earning a bonus for his outstanding work. This positive note soon turned negative. The latter half of the letter brought up the deteriorating conditions of the industry at that time, especially the businesses that Centralab was engaged in. It ended with the stern warning that there were soon going to be some tough times for everyone. In short, business was falling off, and serious belt-tightening would be in order.

It was about this time that Jack decided to see about changing jobs to another company in the semiconductor and miniaturization industry, and he immediately began his search. His interesting resume promptly brought requests for interviews from Texas Instruments, Motorola, and IBM, which resulted in offers from both TI and Motorola. Motorola's offer was for a job that split his time between transistor production duties and working in their miniaturization group. The offer from TI said simply, "We want you to work in our new micro-miniaturization department."

There was an information bulletin issued at Centralab: "J. S. Kilby is resigning May 15 to take a position as Senior Project Engineer with Texas Instruments, Inc. E. Michalak will take over the responsibilities as Product Engineer of the S/C group and will report directly to R.L. Wolff. [signed] R.L. Wolff"

Jack was history at Centralab, but was soon to become a history-maker at TI.

Chapter Twelve

Kilby invents the integrated circuit

Jack starts to work at Texas Instruments and soon has a really good idea

Caution! Technical language ahead! Prepare to ignore!

BY THE MID 1950s, the necessity of "miniaturization" was catching on in the industry. The tyranny of numbers was not just a cute expression. It forecast a dead end for growth in the electronics industry if a fundamental change wasn't forthcoming in the next decade. The transistor had greatly eased the immediate vacuum-tube pain, with huge reductions in size, power consumption, and reliability, but it had neither reduced the parts count nor the interconnections to hook them all together. Different approaches were being pursued, and most of them were sponsored by U.S. military research groups. Varo Manufacturing in Garland, Texas, was working on "integrated microcircuitry," a thin-film approach sponsored by the Army-Navy Instrument Program (ANIP). The "micromodule" idea was being pursued by the Signal Corps and RCA, using a standardized building-block technique. The "DOFL wafer," was a ceramic substrate with recesses, whose program was in the Diamond Ordnance Fuse Laboratory.

In 1957, the TI Semiconductor Development Lab was already up to its PhDs in semiconductor projects when its talented director, Dr. Willis Adcock, the physical chemist who led the silicon transistor breakthrough in 1954, picked up another challenge. Pat Haggerty, TI's far-seeing president, had suggested to Willis that this microminiaturization approach to whip the tyranny of numbers sounded like it might have some legs, and that he should take a look at it. Anyone at TI at that time knew a suggestion from Haggerty was the exact equivalent of a direct order, so Willis mentally stuffed yet another project into his overloaded group.

There had been an announcement in late 1957 that RCA had received a multimillion-dollar contract from the Signal Corps to work on the micromodule approach. RCA had stated that they would welcome partnering with other interested manufacturers. Willis visited RCA and talked to a vice-president about this opportunity, but came back disappointed with the less-than-enthusiastic exchange of ideas. It wasn't until sometime later that he received an official response: TI made NPN transistors and RCA made PNPs, so there was a compatibility problem. Thanks for your offer of help, but no thanks.

So Willis wisely decided to go fishing, engineer fishing, and put out employment ads for engineers with experience in semiconductors and miniaturization. By a stroke of good luck for TI, Jack Kilby of Centralab had also decided to go fishing. Jack had realized that Centralab would not have the resources to take a serious bite out of the developing transistor world, so he cast his bait to two of the industry leaders, IBM and Motorola, and then one more to an upstart in Dallas, Texas Instruments.

When Dr. Adcock received Jack's letter, he remembered him from attendance at some of the later transistor technical conferences. Kilby and Willis met for an interview, and they both liked what they heard. Kilby's only other job prospect was from Motorola, and it was for part-time miniaturization work, while Willis wanted a full-time person with Jack's experience to propose winning government "R&D," research and development, contracts for TI. Things looked very promising except for one small snag—Willis had been told to staff up his potent development group with PhDs, and Jack was one degree short. Willis also knew when to break the rules, and he did. He would later say that hiring Jack was his most important contribution to TI.

Exactly what talents did Kilby bring to this TI team? First, he had two degrees in electrical engineering, and a solid knowledge and practical understanding of electronic circuitry, both digital and analog. Second, he had spent years puzzling over ways to shrink and simplify electronic devices and, in addition to having a thorough understanding of the current approaches, Jack recognized which of these were dead ends: the ideas which had fatal failings, either technical or financial.

The third piece in this package of talent was Jack's rare transistor experience with Centralab. It was a breadth of experience that few engineers in the world had. It was the keystone in his expanding educational experiences that carried him to the invention of an eon. His work had allowed him, caused him, to take in the entire world of semiconductors, limited though it was at the time. In a few years, Jack had gone from absolute zero to the application of transistors in a consumer product. When Jack walked into the Bell Labs transistor symposium, blank notebook in hand, this crucial experience began. It ended with transistor-equipped hearing aids rolling off Centralab's production line. Jack had done it all—conceived and set up the transistor production line and designed the hearing aid circuits. Few engineers ever experience that range of total immersion in a new technology. Jack had the perfect background in education and experience to bring the tyranny of numbers crashing down.

The missing piece that kept the idea of the monolithic circuit from springing forth from Jack's forehead was a place and an atmosphere that would let this happen. And what would that be? A place with the tools and equipment to build semiconductor structures; a capable, supportive, and non-intrusive leader; and, most important of all, a time to quietly think without becoming mired in day-to-day work-place activities. These conditions were well met by Texas Instruments in the form of a small but top-notch experimental lab staffed by talented people, the inimitable Dr. Willis Adcock for a boss, and the summer mass plant vacation that Jack did not qualify for. The last piece of the puzzle clicked into place.

Kilby was coming to work in what was later described as the TI equivalent of the infamous Lockheed "Skunkworks." Jack admitted that when he started, he was virtually self-directed, which reflected favorably on TI's general culture of "take responsibility and do it if you think it's right," and especially favorably on the sanguine leadership of Adcock in the Development Lab.

Within a few weeks of his arrival, Jack was issued his personal Lab Notebook. All TI engineers and scientists were trained to document their experiments and results. It was going to become one of the most famous and valuable books in the world.

LABORATORY NOTEBOOK

In Custody of

Jack Kilby

Date of Issue

June 12, 1958

Property of

TEXAS INSTRUMENTS
INCORPORATED
6000 LEMMON AVENUE DALLAS 9. TEXAS

The following are text and sketches from this notebook:, with my comments in brackets:

Page 1, June 13, 1958

Method for Assembly of Circuit Elements

Many electrical circuit elements can be most economically manufactured in a rod or tubular form. This form has many advantages in processing and material handling. Resistors and tubular capacitors are currently available. Transistors could also be made in this shape, shown by the sketch below....

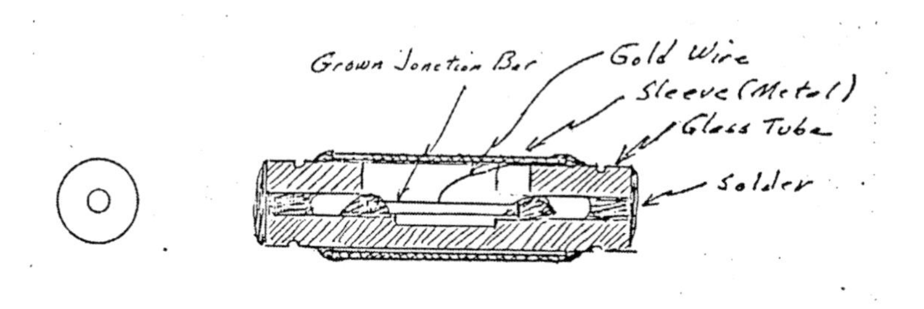

[This is Kilby's first notebook entry, and it is an idea pursuing the micromodule concept that RCA had contracted from the Signal Corps. It is to make all circuit elements physically similar, plug them into a standard "cube" and electrically connect them with printed wiring to the electrical contacts on their ends.]

Such transistors could be readily assembled with other tubular components as shown below:

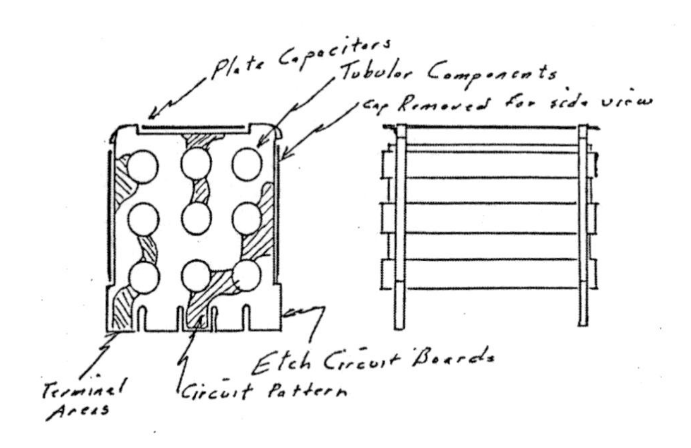

Up to nine tubular components could be dropped into place, by machine if desired,

> [Well, almost all components. The "standardized shape" didn't last long. Some capacitors needed to be flat plates, not tubular.]

Since large values of capacitance can best be obtained in plate form, they have been used in the sides & top of the unit. The connections between components would be made by conductors on the etched circuit boards. One end of each board has been shaped to serve as a lead—that is, the terminal areas could be dropped thru a slot in a mother circuit board...

June 27, 1958

The first working model of the transistor shown on page 1 was completed today. The tubes used were metalized with gold paint by H N Riser [Henry Riser], *and the junction bar* [the transistor element] *assembled by Phil Ferguson* [?]. *The gold metalizing does not tin readily, and the unit is probably not hermetically sealed.*

July 3, 1958

Five more good transistors meeting the 2N308 & 2N309 electrical test specs were completed on July 3. These units used Hanovia silver paint, and appear to be well sealed. They will be used to construct an IF Strip.

July 12, 1958

A complete IF strip, consisting of two stages of amplification and a third stage with a diode detector has been constructed with the circuit below...

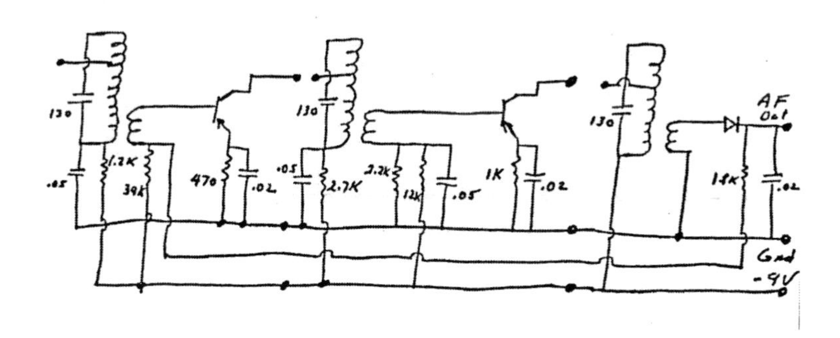

This circuit was developed by Rodger Weber [Roger Webster]. *The three stages when constructed, were assembled on a small mother board for testing. ... the unit was found to have 54 dB electrical gain...*

[The tubular transistors worked well.]

July 13, 1958

The three IF stages were assembled into a small radio chassis... and demonstrated to Willis Adcock, Charles Phipps, Mark Shepherd, and others...

July 22, 1958

...is a Diffused Base transistor structure. This structure uses a small glass or ceramic chip to support the wafer.

[Jack tried a new and better type of transistor in the tubular configuration, but this is his last notebook entry on the micromodule approach. Kilby not only felt the approach was a dead end, but even worse, he had a growing fear that he would be put in charge of drafting a proposal to the Signal Corps—a contract proposal to develop the micromodule approach—his reward for having already cracked the code on tubular transistors. Few things are more repulsive to a creative engineer than writing proposals to government entities, especially when the proposals are at cross-purposes with their own ideas on how the job should, or maybe shouldn't, be approached. Jack was strongly motivated to leap-frog the threatening micromodule ogre with an idea of his own that he'd been kicking around since he'd been at TI. He was going to take full advantage of the TI two-week "mass vacation" in July enjoyed by most employees, which had left the new and cavernous Semiconductor Building on the TI campus refreshingly deserted and peaceful.

He was quoted, "As a new employee, I had no vacation time coming and was left alone to ponder the results of the IF amplifier exercise. The cost analysis gave me my first insight into the cost structure of a semiconductor house." The results were eye-opening—there was no way the micromodule approach could ever make more than a small dent in the cost of electronic circuits. But silicon? Maybe everything on a piece of silicon?

He said later that as he developed this growing vision, he'd sketched out each step with "lots of layers done with colored pencils" to dope out and clarify the new three-dimensional electronic component fabrication and connection system. Jack had added a dimension to the 2D "flat" schematic of circuits—he was going use the up-and-down as well as the round-and-round.]

July 24, 1958

Extreme miniaturization of many electrical circuits could be achieved by making resistors, capacitors and transistors & diodes on a single slice of silicon.

[That didn't take long. Jack laid out the principles of the integrated circuit in the first sentence.]

...If the slice were thin, useful values of resistors could easily be made by attaching ohmic contacts...

...shows that junctions which have a large difference of resistivity and a sharp discontinuity of resistivity at the junction have a capacitance of...

The following circuit elements could be made on a single slice:

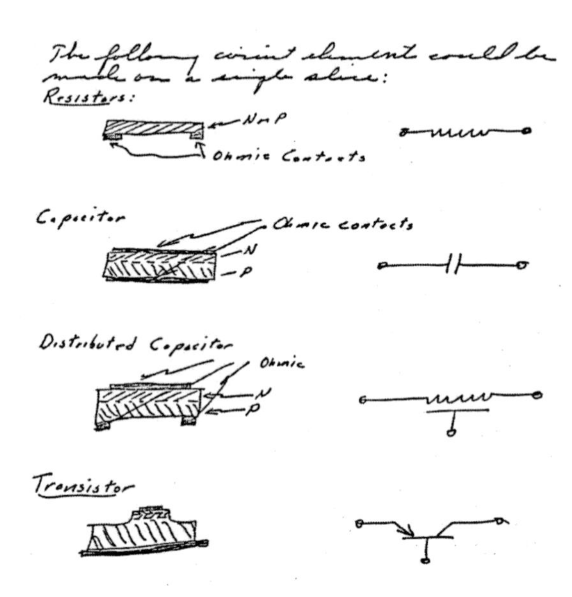

[Resistor, Capacitor, Distributed capacitor, Transistor]

A process which might be used for fabricating a circuit on a single crystal slice of silicon is outlined below.

First, a slice of the proper size would be double diffused...

[Jack identifies each production process step, with sketches of the proposed device as components are added.]

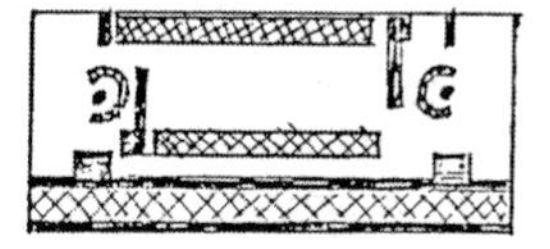

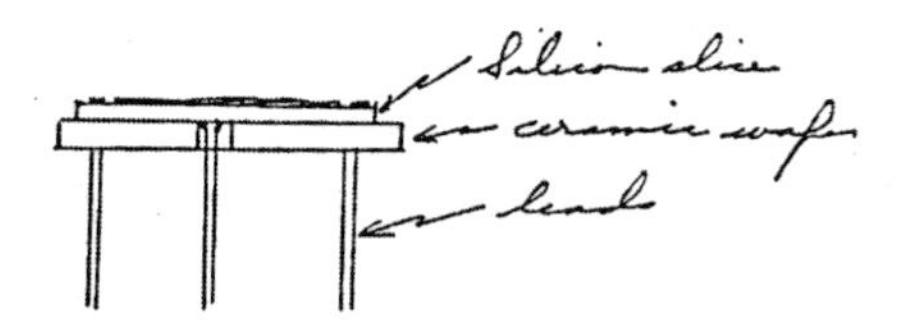

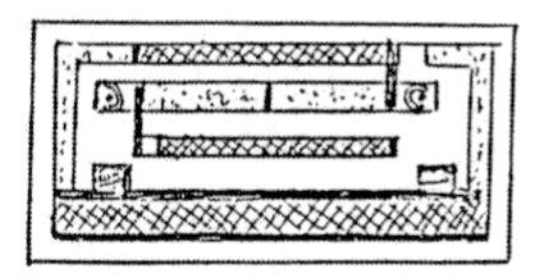

[Six pages later]

This unit would have all of the circuit elements for a multivibrator.

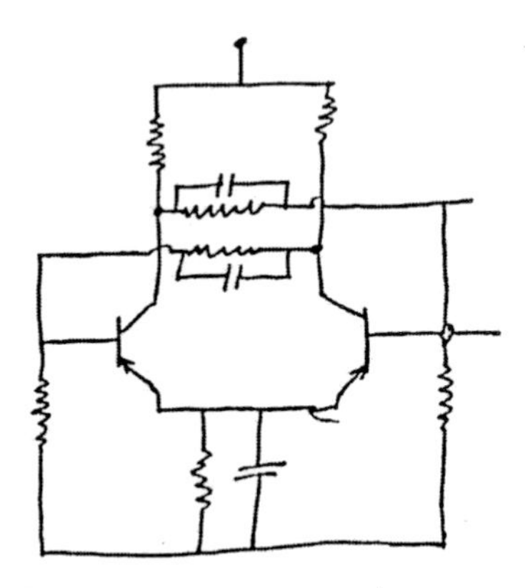

[signed] *J S Kilby July 24, 1958*

[TI's mass vacation was over, and the clamor and flurry of urgent activities of the development lab had returned. But his "quiet time" mission had been accomplished, and Jack spent a busy and productive July 24th putting his new concepts into the notebook. These were shown and explained to Willis Adcock and Pat Haggerty. Kilby said Willis thought it looked interesting, but had some reservations. Pat Haggerty, who seemingly could out-X-ray-vision Superman, saw something in Jack's sketches that excited him.

Willis, proceeding cautiously, asked Jack to build the individual electronic components separately out of silicon, and wire them together and see if the circuit would work. Jack agreed.

Jack did the math and tests necessary for the design, then had the individual components needed for a multivibrator built in the lab using slivers of a silicon wafer.]

August 28, 1958

A multivibrator circuit was selected to show the feasibility of all silicon circuits...

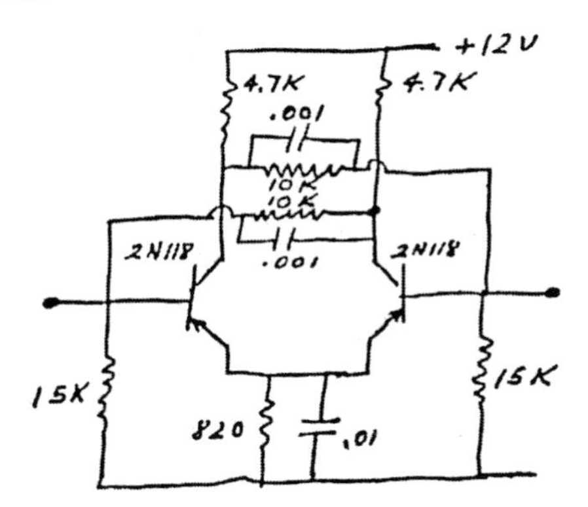

Silicon elements were prepared for the circuit above. ... This unit was assembled & tested by feeding square waves thru coupling capacitors into both bases. Unit followed well up to about 20 Kc.

[It worked, as Jack knew it would. The first small hurdle was over, and next would come the ultimate hurdle—put all the components on a single piece of silicon and see if it would still work. Jack was convinced, "If you can make all the parts from the same material, you can make them from the same piece." Willis told Jack to go for it.

Jack prudently decided less was more for the big jump to a monolithic circuit, and opted to use a single-transistor circuit, a phase-shift oscillator, instead of the more complex multivibrator. It was simpler, but would serve to identify any problems. Jack also elected to use germanium instead of silicon because of the relative ease of doing the necessary process steps in the lab. Again, it would prove or disprove the idea as well as the more difficult-to-work silicon.

Jack sketched up his design and sent it to Tom Yeargan and Pat Harbrecht, two skillful lab technicians, to build. In fact, said Jack, build a bunch of them, since luck would play a part as it does in any semiconductor manufacturing. It took at least several days to build the rough devices, and it can only be imagined that even the imperturbable Kilby might have been a little excited while waiting to see the outcome of this embryonic experiment.

But first, Willis dropped by with an interruption, as bosses sometimes do. Would Jack see if a person could make a phonograph pickup, used to play vinyl "records," the CD of the day, by vibrating a grown-junction transistor bar? Jack called King's X on his project, and on September 9 and 10 fiddled with making a phono pickup.]

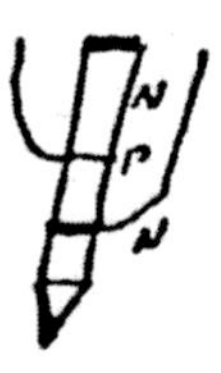

[It worked, kind of. Then Jack went back to the lab to see if his devices were ready.

It could be surmised with some sense of certainty that the six monolithic phase-shift oscillators that Yeargan and Harbrecht built were delivered to Jack on September 12, 1958. Could any engineer have waited until the next day to see if they worked?]

September 12, 1958

A wafer of germanium has been prepared as shown...

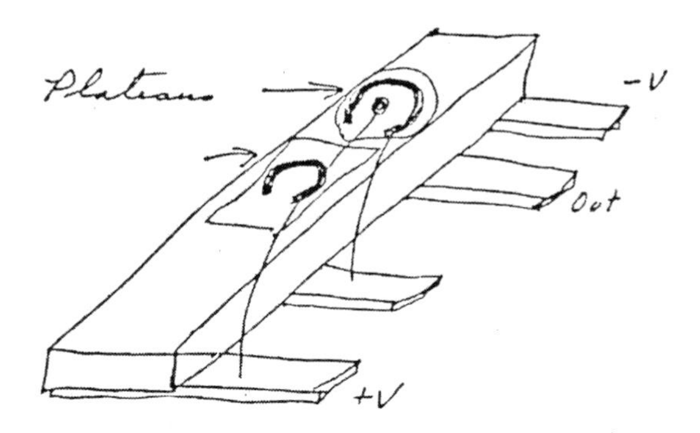

When 10 volts were applied (1000 ohm series current limiting resistor), the unit oscillated at about 1.3 Mc, amplitude about 0.2 v pp, This test was witnessed by W. A. Adcock, Bob Pritchard, Mark Shepherd, and others.

J S Kilby
September 12, 1958

[When asked in an interview years later about his feelings at that time, Kilby said, "It was somewhere between pleasure and elation." Maybe he wouldn't have to write that government proposal after all.]

Chapter Thirteen

The IC slowly comes to life

After Kilby's inspiration came the perspiration, and there was plenty of it from everyone at TI.

LOOKING BACK THROUGH HISTORY, the invention of the integrated circuit by Kilby in 1958 now seems to have been the easy part. It was more akin to giving birth to a tiny premature baby than a squalling red-faced eight-pounder; a preemie afflicted with a variety of ailments which did not bode well for its immediate future, much less growing to a robust adult. It was a hard sell, and Jack Kilby and his marketing partner, Charles Phipps, had to take on the technical world and change its mind.

The first official press announcement of the idea that was to change the world was made by Pat Haggerty of Texas Instruments in March, 1959, during the big IRE, the Institute of Radio Engineers, trade show in New York City. The announcement of TI's miraculous contraption, a tiny germanium flip-flop circuit, was made to the press at the New York Athletic Club. The trade show newspaper headlined, "TI Shows Match-head Solid Circuit – Tiny Device To Make Computers Smaller." Even with the promise of smaller computers (which few people had or cared about at that time), the announcement came out a poor second to Fairchild Semiconductor's release of a new

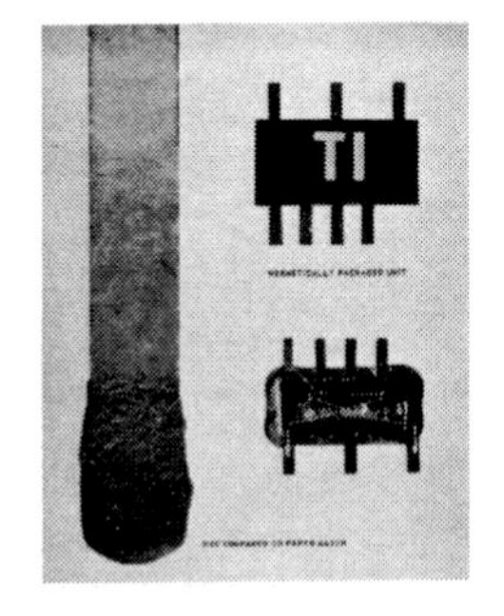

transistor. The difference? The difference between a real transistor and a laboratory curiosity.

Jack had entered the punishment phase of being an inventor. His new "curiosity" caught the miniaturization groups by surprise, and it was not a pleasant surprise. These were the groups funded by military development money—micromodule, molecular electronics, and thin film technology—and they were not interested in an outsider jumping into their private sandpile and messing things up. Kilby became a frequent attendee and presenter at a continuing series of panel discussions and seminars on the state of miniaturization. Some of the offended labs sent "truth squad" scientists and engineers to ask tough questions about TI's new invention as their contributions to the programs. One major manufacturer sent the same two representatives time after time.

The points of attack to discredit the TI claims were the thermal problems of a super-compact array of circuit elements, the lack of "yield," which is a measure of how many devices actually work when they come off the production line, and the shotgun scattering of the performance parameters, any one of which could be a deal-breaker. Charles Phipps reported, "Jack did not particularly like to go to these talks or panels, but he did and probably made over half of them, and Dick Lee and myself would go to the others." What's not to like about having the entire audience boo down your new invention? Kilby, ever the understater, has been quoted, "We were the source of considerable amusement."

Meanwhile, back in the laboratories at TI, the new baby was taking nourishment and gaining weight. The Signal Corps had shown some interest in tying Kilby's new idea into their own family, the micromodule program, and was investigating the possibilities of a technical merger. But by the time the compatibility issues had been worked out, it was too late. The micromodule program had lost favor and was on the downhill slide to oblivion. In any case, the radical TI approach to miniaturization had stirred up a debate within the Air Force. Should TI get some of the budget money scheduled for

Molecular Electronics development? Not surprisingly, the Molecular Electronics development groups were against it, saying that the TI concept did not meet the requirements of Molecular Electronics. However, a small but technically strong group in the Air Force felt that the TI schema had merit—enough merit to spend some money to help it along. Our hero, R. D. Alberts of the Wright Air Development Center, sent a request for a proposal to TI on March 5, 1959. Dr. Willis Adcock and J. W. "Jim" Lacy responded with an impressive document titled, "Proposal – Semiconductor Solid Circuits," in an equally impressive short time. It was dated March 18, 1959. This resulted in the first modest Air Force contract to fund the further development of the concept and manufacturing techniques of the Solid Circuits. The cavalry had arrived and this time it was the U.S. Air Force. Jack's little newborn was beginning to put on some serious poundage. Haggerty later stated that the progress of the integrated circuit "owes a great deal to the encouragement, guidance, and financial support by the Air Force, beginning in early 1959."

The next step was the announcement of an integrated circuit, at that time called by the TI copyrighted name, "Solid Circuit," at the spring Institute of Radio Engineers, the IRE, show of 1960. The first circuit was a TI Type 502 flip-flop containing 16 circuit elements, at the bargain price of only $450. The electronic revolution had begun in a very small and very expensive way. But it had officially started.

Teaming up with the Air Force had a two-pronged benefit for Solid Circuits. The Air Force was funding some of the development, and the Air Force needed some new electronic gadgets that might be built from them. What a revolutionary idea! Someone could actually build a useful product with this baby? Looks like the baby has just pulled up and is looking around the room while holding on to the coffee table. You don't suppose…?

In 1961, TI delivered an astonishingly compact digital computer to the Air Force. It had been built with their help, and it contained 587 circuit elements, Solid Circuits, if you will. The design team was led by TI's talented Harvey Cragon, and the chips were furnished by

Kilby's gang. It was an extremely ambitious undertaking for the day. The previous cat-calls of "laboratory curiosity" from the opponents weren't heard, as TI had simultaneously announced the availability of a family of Solid Circuits—six digital circuit devices. The Series 51 Solid Circuits were in production. Jack's baby had just run across the room and out the door.

This successful Air Force computer led to the 1962 TI contract to design and build a family of 22 Solid Circuits for the Minuteman II missile guidance system. Near the end of 1964, the first model of the missile was successfully flight tested. The electronics contained 2,200 of the tiny circuits. Jack and his cohorts were no longer furnishing the amusement. The competition was too busy trying to jump on the bandwagon to even smile.

Jack S. Kilby and his group of scientific stalwarts had battled through six years of backbreaking technical and heartbreaking political struggles. They had brought the integrated circuit to a viable state and proved Kilby right—you really could put all those electronic thingies on a single piece of silicon.

Meanwhile, Pat Haggerty, TI's fearless leader, had been casting about for a product that could be made using Solid Circuits. A product that real people could use, not just the Air Force. Pat had been responsible for letting the transistor out of the bag in 1954 when he brute-forced the development of the Regency TR-1, the world's first pocket transistor radio. This product soon developed a huge market for the underappreciated transistor and, as is always the case, brought on big production which brought lower prices, which brought on more sales. Within a few years it seemed everybody in the world had a "transistor," as the little radios were called. TI needed to do something similar with the Solid Circuit, now being called the integrated circuit, or IC, and later, just "the chip." Haggerty thought it was time to spring this gadget on the public.

Chapter Fourteen

The pocket calculator

What can add, subtract, multiply, and divide, but won't fit in your pocket? (See answer below.)

KILBY REMEMBERED THE FLIGHT to New York City with Pat Haggerty on a business trip when Pat shared his new product musings with him. Pat had come up with three possible Solid Circuit products for Jack to think about. One was a lipstick-sized voice recorder, another was a pocket calculator that would add, subtract, multiply, and divide, and the third idea was so strange Jack claimed that he didn't remember what it was. But the calculator had caught his attention.

Haggerty's musing carried the same horsepower as his suggestions—go do something about it. This significant musing and Kilby's subsequent reaction can be dated in a odd way. In the SMU Kilby Archives is an unmarked folder containing eight sales brochures for breadbox-sized mechanical and electronic desk calculators, state of the art for the times. The brochures are dated 1964, and the cheapest one, a mechanical job, was over $1,000. Very encouraging, indeed.

Texas Instruments set up a small but technically potent secret project to develop a pocket, or more nearly, hand-held calculator that could do the four basic arithmetic functions and display the answer to the person whose hand was involved. In addition to Jack, who of course would have the monumental task of developing the necessary integrated circuits, the team had Jerry Merryman. The selection of

Merryman for the digital design of a calculator in 1964, was both an easy and an inspired choice. Kilby said later he doubted if anyone else in TI could have done what Jerry did. Jerry was, and still is, the ultimate polymath—his range of knowledge is both vast and immensely entertaining. The third member of this team was Jim Van Tassel, a versatile scientist, who was to do the design of the peripheral bits and pieces of the gadget. This secret project was named "Cal Tech." No one could have guessed what it was from its code name.

Kilby was busy. Jack had been the manager of the Series 52 and Series 53 integrated circuits in 1963, which must have been a killer job. In December 1964, he had been promoted to Program Manager of SRDL, the Semiconductor Research and Development Laboratories. A brief month later, on the first day of 1965, he was bumped up to Deputy Director of SRDL. In this demanding position, he took on yet another job—Haggerty wanted his hand-held calculator and he wanted Kilby to make the integrated circuits for it. Things would never be the same as they had been during Jack's first glorious months of solitary and productive work at TI.

Pat Haggerty had a famous saying, "The reward for a job well done is a harder job," and Kilby had felt the "benefits" of this philosophy many times.

Years later, Mark Shepherd volunteered the information that one of the few things he would do differently if he were to relive his life as president of TI, would be to let people do what they did best. Rewarding creativity by promotion into management was not necessarily a good thing. Kilby was on the receiving end of both philosophies. He had just been rewarded with a harder management job. TI would later establish the TI Fellows Program, making it possible for exceptionally creative employees to continue doing what they did best, and progress up the salary ladder without having to be supervisors.

Merryman pulled off one of his best engineering miracles and put the circuit diagrams of a four-function calculator on paper in a non-stop solo design orgy of three days and nights. The Regency radio had 4 transistors in 1954, and Merryman's new calculator 4,000. The secret Project Cal Tech was off and running. The team had a working model by December 1966 and filed a patent, "Miniature Electronic Calculator," the following year.

The development of the necessary integrated circuits lagged years behind the circuit design, but the sophistication of the manufacturing technology rose slowly and steadily under the guidance of Kilby's talented and dedicated team. After supplying early calculator chips to Bomar and Canon for two years, TI announced their own calculator in 1972, the TI-2500 Datamath. It was an elegantly styled addition to the marketplace at the eye-popping price of $149. This was the result of TIers Gary Boone and Mike Cochran cracking the code and achieving the ultimate goal—the single-chip calculator.

Jack's pioneering work was done, and the people of the world were soon going to be able to put down their transistor radios and slide rules, and pick up a calculator: a hand-held calculator with one chip of silicon that held 5,000 transistors.

Chapter Fifteen

Goodbye, TI

"Independent Inventor" trumps "Assistant Vice President of Texas Instruments"

JACK'S SUCCESS AT TI was taking him in the wrong direction. First, he was promoted to Deputy Director of the Semiconductor Research and Development Labs, and then, in September 1968, made an Assistant Vice-president of Texas Instruments. How much can a guy stand? What next? He was beginning to begrudge the time he spent in what he described as an advisory role, the percent of which had been gradually increasing over the years. Kilby missed his participation in the creative level of projects, and was not satisfied with keeping track of projects and going to meetings—meetings to hear presentations from the projects he was responsible for, and meetings for him to pass on the status of the projects to his bosses.

He would have agreed with Mark Shepherd's post-presidential change of heart to keep a person doing what he was good at. Jack was good at inventing but was now shoving numbers around and making "foils" (TI's name for projection transparencies) for presentations. It would be safe to say he was making a lot of foils.

An early clue to Jack's inner conflict might be found in a 1964 memo to S. M. "Mack" Mims, TI's senior patent attorney. Jack was working on an invention, a "teaching machine," that was not related to his work at TI, and he wanted a release from TI so he could get it patented "outside." It was a standard procedure at that time to get a signed agreement from all incoming employees that anything they invented belonged to the company, whether they did it at home in their easy chair or at their Steelcase desk at the office. In actual practice, TI was willing to sign the rights of an idea back to the inventor if it was "outside TI's field of endeavor."

But it seems that TI thought they might go into the teaching business someday, which did indeed come to pass in the late 1970s with the Speak and Spell, The Little Professor, and other "learning" boxes designed to help the school kids with their lessons. A patent application for "Teaching Machine," invented by J.S. Kilby, was filed in 1966. It resulted in a U.S. patent being issued in 1968 that was assigned to TI, not J.S. Kilby.

A few years later, after Jack had taken leave of absence from TI, he tried it again. A patent application for "Teaching System," invented by J.S. Kilby, was first filed in 1971. This patent was issued in 1977, and was assigned to J.S. Kilby. Jack was inventing, whether he was at TI or not. He couldn't help it.

The idea of "inventing" is a nebulous concept. How do you decide to invent something? Maybe more to the point, *why* do you decide to invent something? To the born engineer, as Jack seemed to be, it was more of a question of how do you *not* invent something? It would be folly to speculate on the workings of Jack's brain, and even more so to try to pin-point the cause of Jack's innate curiosity which led to so many original ideas. Take a look in "Patents," at the end of the book, if you want to see the variety and scope of Jack's inventive nature. Curiosity seems to be one of the traits that an inventor would need. But what else?

We get another glimpse into Jack's creative processes if we hear his response on one occasion to a question about being an inventor. He said, "If you want your kids to grow up to be inventors, read them fairy tales." Maybe that was the missing term in Jack's equation of success: continual curiosity plus a spirited imagination equals creativity.

Jack wanted to strike out on his own as an independent inventor, and he was granted an extended leave of absence from Texas Instruments on October 31, 1970. Maybe he danced out the door and clicked his heels as he might have done in grade school at vacation time, singing as he went,

> *"No more meetings,*
> *No more books.*
> *No more bosses'*
> *Dirty looks"*

and then slowly disappearing like the Cheshire cat, his smile being the last thing to fade away.

Or maybe not.

Chapter Sixteen

Independent inventor

The sweet freedom to pursue any and all ideas, and an unlikely one brings Jack back to Mother TI.

KILBY BRUSHED OFF the remnants of the TI hustle and bustle, and went into business for himself. The first of Jack's new engineering notebooks was identified as "Vol. 101 J.S. Kilby Nov 25, 1970," in the same distinctive binding that had proclaimed "June 13, 1958" in his first Texas Instruments notebook.

Jack worked hard as an inventor. In the next seven years there are 31 entries relating to different ideas and inventions—some brief, some lengthy. Some are recognizable by patents that were later issued. Some were not pursued to the stage of a patent application. Some were serious technical forays, another appeared to be an engineer's response to a personal annoyance. Some of the patents were sold to other companies, most were not.

There was an entry dated April 26, 1974 in Volume 102 that was destined to keep Jack hopping for quite a while. It was an idea for a solar-powered hydrogen generator. Some time in the future you might put this thing on your roof on a sunny day, and it would generate hydrogen gas. Hydrogen is good. It can be burned to cook your food or heat your house. It could power your hydrogen-fueled automobile, and, if you put it through a fuel cell (perhaps outside your back door next to the garbage can), it could make electricity for your home.

The next entry concerning this novel idea was August 21. This entry was 27 pages long. Jack had launched himself into a big one. He hired consultants to do laboratory work that he was not equipped to do, and a key entry stands out in Volume 103—there was a successful test of the solar hydrogen generator at College Station, Texas, on January 13, 1975.

Jack soon took his brainchild on the road, back to the company he knew best—Texas Instruments—and on April 5, 1975, showed his solar energy system to J. Fred Bucy, Chief Operating Officer. On June 30, he showed it to the TI board of directors, including Pat Haggerty, the chairman. Jack was making good progress. Haggerty, Jack noted, was pretty excited about it.

TI was convinced that the idea had promise, and decided to team up with Kilby and investigate this unique method of turning sunlight into hydrogen or, if you prefer, into electricity using fuel cells. Jack suggested to the powers at TI that Pete Johnson would be a good guy to lead this project, and they agreed. Pete recalls going to Jack's office and sitting down with him for a couple of hours and, "We kind of firmed it up right there." Project Illinois, in honor of Jack's alma mater, had begun at Texas Instruments.

Pete Johnson was the project manager and Kilby the consultant. The project gradually staffed up over several years, with a top personnel count of 85 to 90 persons. Of these, about half were "exempt," meaning on salary. Of these exempts, about half were PhDs, which made this a very potent research organization. Pete explained, "We were rich in electrochemists, since there was a lot of electrochemistry involved."

After several years of the initial spadework on Project Illinois, TI negotiated and won a cost-sharing program with the Department of Energy's solar group in Golden, Colorado. It was the first cost-sharing program DOE had ever sponsored. This generous program between TI and the DOE lasted for four years. The program ended with a final exam—TI was to build and test a "production-quality" system. A unit was built, and it not only passed all the tests, it did it within budget and on time.

The next obvious step was the big one: build a pilot plant and go into business selling residential-sized solar converters—let the sunshine in and make some energy to run your home. The system showed good numbers in the efficiency column, and the cost was estimated to allow pay-off in a reasonable time. It was a unique product for the consumer market, the likes of which no one had seen since the days of the 1930s when a Delco wind-driven generator spun on top of the barn, charging a 32-volt storage battery to run farmhouse lights and radio. Free energy, kind of.

A rough estimate for the pilot plant was a nice round hundred-million bucks, which would have been a big pill for TI to swallow, even in the best of times. And these was not the best of times for TI. Mark Shepherd, chairman and CEO, the big boss, said if you want to go into the solar energy business, go find a partner to team up with—maybe someone already in the energy business, and preferably with lots of money. So Jack and Pete, sometimes in the company of George Heilmeier, TI Chief Technical Officer, hit the road and the sky. They crisscrossed the United States and several foreign countries, but found no takers for the program in nearly a year of trying. In the meantime, TI had begun making noises of returning to what they did best, which was making semiconductors, not solar energy systems. Lacking a really rich sugar-daddy, it was the end of the line for Project Illinois. In September of 1983, TI pulled the plug and told Pete to disband and place his team members in other projects at TI. Eight years of Jack's life and TI's efforts had ended up a great big and extremely disappointing zero.

Even with the heartbreak ending, Pete Johnson said that this was "the best eight years of my time at TI." His next sentence told why, as he spoke of working with Jack Kilby, "We had a wonderful relationship."

Meanwhile, back at TI, Jules Levine, a scientist who managed one of the technical groups within the rapidly dissolving Project Illinois, had an idea. He thought that the key idea of the spherical silicon solar cells could be split out of the hydrogen generating system and used to generate solar electricity without the complexity and hazards of the chemical solutions. Kind of a dry solar system, and it could be an efficient way of going directly from solar energy to electricity if you didn't want the hydrogen. He was also a good salesman, as TI set up a small group to give the idea a try. In about two years the small group had developed a promising solar energy system using the unique feature from Project Illinois. But TI had already made the decision to get out of divergent businesses, no matter how promising they might look, and the project was disbanded.

But the rising green of solar energy brought renewed interest in TI's and Kilby's patents, and in 1995 a company bought the rights to the dry solar system technology. This company had seen something promising, too, and has since turned at least part of Jack's original idea into a useful product. Jack's idea is now making solar electricity.

The failure of the TI solar program to survive to completion was a tough blow for Jack, who had never lost confidence in the idea or the path of the program. Its demise was like the last straw in a series of personal family misfortunes. Jack's father had died in 1971, and his mother later moved to Dallas to live her declining years near Jack's sister, Jane. She died in April of 1980. Then, after a long illness, Jack's beloved wife Barbara died in November, 1981.

When the solar project ended in 1983, Jack was 60 years old, and hardly ready to retire. Maybe Jack felt he could slow down, and from all appearances, he did. The steady stream of inventions that he logged in his notebooks had ceased during the latter part of the solar energy program, and never started again. TI's abandonment of the solar project brought Jack to the point of telling them a final goodbye, as he may have felt that his work with the firm that brought him world-wide acclaim was finished. Jack St. Clair Kilby officially terminated his Leave Of Absence from Texas Instruments and retired November 1, 1983. *Sic transit gloria mundi*, indeed.

Jack had been awarded the position of Distinguished Professor of Electrical Engineering at Texas A&M in 1978, and in 1984 he called it quits there also.

Not that Jack was going to lack anything to do. He was still on the boards and advisory committees of several companies and government agencies. He occasionally lectured and did consulting work. He was also the recipient, at a steady rate of one every couple of years, of prestigious national and international awards. And, of course, he was still being interviewed by a variety of news services, which had started years before as his quiet fame had spread. But Jack the Giant Killer Inventor had turned down the wick of his flame, and seemed to be content to hang up his slide rule.

Chapter Seventeen

The Kilby family

L to R: Ann, Barbara, Janet, and Jack

The Kilby family vacationing in their favorite place, Colorado, 1963

JACK'S INVENTION of the integrated circuit has been documented with about the same fervor as the Wright brother's first airplane flight, and appropriately so. Both of those events were inevitable, but both startled the world when they occurred, although the timing between the event and the startling was vastly different. Everybody instantly understood the miracle of a machine that flew, because they could see it happening. But they couldn't foresee the Wright's canvas and wire contraption turning into the SST. Few engineers and scientists understood the integrated circuit when Jack gave birth to it, and it was going to be a very long time before even the experts could faintly conceive of the tidal wave it was going to create.

The part of Jack's life that is not well documented are the hours away from his work and the technical community—at home with his family. And, not too surprisingly for the few who knew Jack well, his time by himself. People that he worked with for years knew very little about Jack and his home life—his hobbies and his interests away from the high-pressure world of high technology. Jack was a great inventor and, in his other life, his quieter life, his pursuits were just as thorough, methodical, absorbing, and frequently solitary.

This lesser-known side of Jack began with a blind date while he was hard at work at the University of Illinois, in his mad dash of the last two years of his electrical engineering degree. Maybe he recognized that all work and no play made Jack a... well, you know. His blind date was with Barbara Louise Annegers, also a student at the University, and one year behind Jack. It had been set up by a mutual friend who recognized their common interest in being tall.

Barbara was from Galesburg, Illinois, daughter of Mr. and Mrs. J. H. Annegers. She had attended Galesburg High School, was in the Thespian Club, and had a role in "You Can't Take it With You" her sophomore year.

Details of Jack and Barbara's dating and courtship are scarce, but Jack's younger sister, Jane, was later to spill a few beans: "On weekends they would go to the student activities, drink beer with the group, and go to dances, but they didn't dance." They became engaged in February, 1948, while Jack was working for Centralab in Milwaukee, and Barbara was finishing her senior year in Urbana. If Jack had been able to drive a souped-up Model A Ford from Kansas to Florida as a teenager, the 200+ miles between Milwaukee and Urbana must have seemed a short haul.

Barbara finished college and received her degree in speech therapy, and she and Jack were married June 27, 1948, in Galesburg, Illinois. Jane continued her story: "She had a couple of bridesmaids, and Jack had a friend from college to be his best man." May I present Mr. and Mrs. Jack Kilby?

After a honeymoon to a resort in Wisconsin on the shores of Lake Michigan, they returned to begin married life in Milwaukee where they lived while Jack worked at Centralab for ten-plus years. During this busy decade, the Kilbys began their family. They had two daughters, Ann and Janet, and a son, John St. Clair, who was sadly lost to SIDS, sudden infant death syndrome.

They built a house and began putting down their family roots. But it was not to last. Jack's need to move ahead technically in his career led to a big and disruptive move to Dallas, Texas. Dallas was a thousand miles and a world away from their home in Wisconsin.

They settled into an apartment on Hillcrest Road, just north of Northwest Highway. Ann Kilby has vivid memories of the place and of her first encounter with a tarantula. Howdy, podner! Maybe it was a Texas version of the Welcome Wagon. Jack began his new assignment as Senior Project Engineer a few miles away at the new Texas Instruments campus on the north edge of Dallas. His desk was in the Development Engineering Group on the ground floor of the brand-new eye-catching 300,000 square-foot Semiconductor Building. Jack had been working there a month before the official building dedication was held on June 23, 1958.

They soon began looking around for a house to call home, and found a suitable one closer to the TI campus. This house became the home for Jack and Barbara for the rest of their lives.

The house was originally a comfortable fit for the four Kilbys, but it soon began to show signs of crowding. Both Jack and Barbara developed hobbies that called for more elbow room. Jack had ramped up his photographic interests, and by 1966 he was getting serious and making plans for a darkroom. Barbara had been silversmithing and making jewelry, using the craft classroom facilities, but had now moved into ceramics and wanted her own creative space.

Plans were drawn up for additional workshop space in the form of a room with a small bath over the existing garage. Then the construction of this nearly 600 square-foot addition stalled out, as the building permit person of the City of Dallas pointed out that there was a "one-story covenant" for the neighborhood, and that it just wouldn't do to stack another room on top. Jack disagreed with this interpretation of the rules and, although we may never know exactly, it appears that he was able to 'splain it to them satisfactorily. A current look at the tax records may show the line of reasoning—this room over the garage did not make the house a two-story house at all. It made it a 1½ story house, and, shucks, that's close enough to one story.

The hobby room was built as planned. According to Ann Kilby, the tiny bathroom may have been to city code, but it sure wasn't to the tall people code.

Chapter Eighteen

Silver and clay

Barbara did it all, and did it beautifully.

BARBARA ANNEGERS KILBY had a built-in creative streak that led to a lifetime participation in art. Art done with the hands—up close and personal—painting, silver work, and ceramics. She had an urge to make things, beautiful things, and she did.

She began unleashing her creativity while the family was in Milwaukee, experimenting with charcoal sketching and progressing into watercolors. The disruptive move to Dallas derailed Barbara's hobby for a while, but after settling in she soon found a new artistic love—working in silver. After studying with a teacher for several years, she learned both jewelry making and silversmithing. Her jewelry was generally simple and with geometric designs. She fashioned a varieties of rings, pendants, and chains, mostly in silver with oxide touches. These pieces were not for sale, but were made for the family. Ann Kilby has a silver Snoopy that her mother made for her when she was about 14 that she cherishes today. Janet also has a unique piece of silverwork from her mother. It's a long, rectangular matchbox shaped like, of all things, an integrated circuit, a microchip. She had obviously combined artistic forces with Jack.

Barbara also made an exquisite sterling teapot and creamer set at the peak of her silversmithing activity. Many hobby jewelry makers could come up with a Snoopy for their daughters, but it's quite another thing to craft a teapot and creamer from flat sheets of sterling silver. The skill level is a dozen times higher and very few students have the patience and aptitude to ever carry their hobby to that level. Barbara did serious silver work. Janet remembers it well, "I remember the sound of the pounding. There was a lot of pounding."

Over the years, Barbara withdrew from her silver work and concentrated on her growing ceramic skills. The new workroom had finally afforded her space for her creative work— lots of room for an electric potter's wheel and supplies. An electric kiln was installed in the garage, and Barbara was now self-sufficient in her pursuit of pottery-making.

Barbara threw and built pieces mostly with earthenware clay, but occasionally would experiment with raku pottery in the classes she took at SMU. She studied under several local artists and was continually experimenting with glazes. Her daughter Ann said she loved to watch the opening of her mother's kiln after a firing, since the glazes "did such unexpected things."

One piece that Barbara sculpted showed her love of the glazes—a melon-sized stoppered container, hand-formed in an unusual and almost sensuous shape. It was glazed with pale cloudy horizontal bands—jet-black at the bottom, grayish as the shape rises. It then fades into light gray which changes into a cream. The cream darkens as it moved over the curves at the top where it ends as a delicate pinkish orange. An exquisite example of glazing virtuosity.

Barbara was a commercial success with her ceramic art. Her work was carried by several galleries. She crafted and fired lovely but practical pieces—plates, dishes, and mugs. She derived a great deal of pleasure, as related by both Janet and Ann, of selling the works that she created, perhaps as a kind of testament to her worth as an artist. Barbara was very talented, and enjoyed her creativity as a solitary activity that did not revolve around her husband and children.

She died of cancer in November, 1981, leaving a pair of hand-built figures that were never finished. Her studio remained untouched after her death.

Chapter Nineteen

Jack and his camera

IT WAS A DARK AND STORMY NIGHT in 1971 when Jim Fischer accompanied his friend Jack Kilby to a photographic exhibition and judging. Jack had entered a photograph in the 33rd annual KINSA, the prestigious Kodak International Newspaper Snapshot Awards, sponsored in Dallas by the Dallas Morning News. It was Fischer's first time to see the photographic skills of Kilby and other local photographers, and he was impressed by the works lining the walls. The visiting judge critiqued the photos in turn, and finally stopped in front of the photograph entered by Kilby—a close up image of a rag-tag musician whose trombone was reflected in his sunglasses. The judge was ecstatic about the photo, and raved on at some length. He finally stopped and turned to the audience and addressed Jack, "Mr. Kilby, I've never heard your name before, and I've never seen your work before. But I can safely say that we're going to hear a lot about you in the future."

IT IS THE RECOLLECTION of Jack's daughters that his life-long love of photography could have begun as the result of his father's similar interest. This makes some sense, in that Jack's connection with cameras began early and stayed late. There are indications that Jack had a darkroom of sorts in his high school years, if not junior high. The high school annual from Great Bend, Kansas, lists Jack as a member of the Photography Club and on the staff of the yearbook. His yearbook staff "assistant photographer" activity carried over into his beginning years in college at the University of Illinois.

Not surprisingly, Jack went off to the war with not one, but two cameras. Wonder of wonders, the negatives of the films and the color slides he shot in India in the early days of his service in 1944 have survived in good condition.

One of the cameras he took with him was a Kodak Bantam (or possibly a Bantam Special), a compact camera that used a special film, Kodak 828. Jack shot black and white film with this camera.

His other camera was an unknown 35mm model that Jack used to shoot Kodachrome slides. He frequently shot pictures of the same scene with both cameras, a sure sign of a camera-happy photographer. Historically, the photographs are a treasure, showing locals erecting the bamboo and thatch buildings that Jack's OSS group were to occupy in the hamlet of Chabua, India. They also captured the image of the surroundings and the people who lived there.

One photograph, missing from the black and white negative collection, was saved only as a print. It is a snapshot of Jack sitting on one side of a dining table in a semi-open-air bamboo-built eating room, for want of better words. At first glance, it looks like it might have been taken in a prisoner of war camp, as Jack is dining without his shirt (as he was in several of the shots that he appeared in) and he looked like he was sorely in need of a solid meal. He wasn't, of course, because that's just the way lanky Jack looked when he was 21. You might remember seeing the photo at the end of Chapter 8, "Kilby goes to war."

Time passed, and Jack was out of the jungle and working at Texas Instruments. In 1965 or 1966, Jack bought a Model 500C Hasselblad, the ultimate professional camera. The Hasselblad is a very expensive camera, but if you're going to take photography seriously, really seriously, you need one. And Jack needed one or he wouldn't have bought it. Jack knew, from many facets of his engineering life, that a person's tools make a difference, and it's appropriate to buy the best tools if you want to be the best.

He joined the Dallas Camera Club in 1967. About this time he also began entering photographic contests. He sent entries to events all over the United States. The photograph below that captured the raising of the steel girders on the old Dallas Cowboy football stadium in Irving, was entered in the Central Washington Photographic Salon in Yakima, Washington; the Mississippi Valley Salon in St. Louis, Missouri; the Wilmington International Exhibition of Photography in Delaware; and the San Antonio International photo salon in Texas. The back of this print looks like the stickers on the suitcase of a traveling salesman.

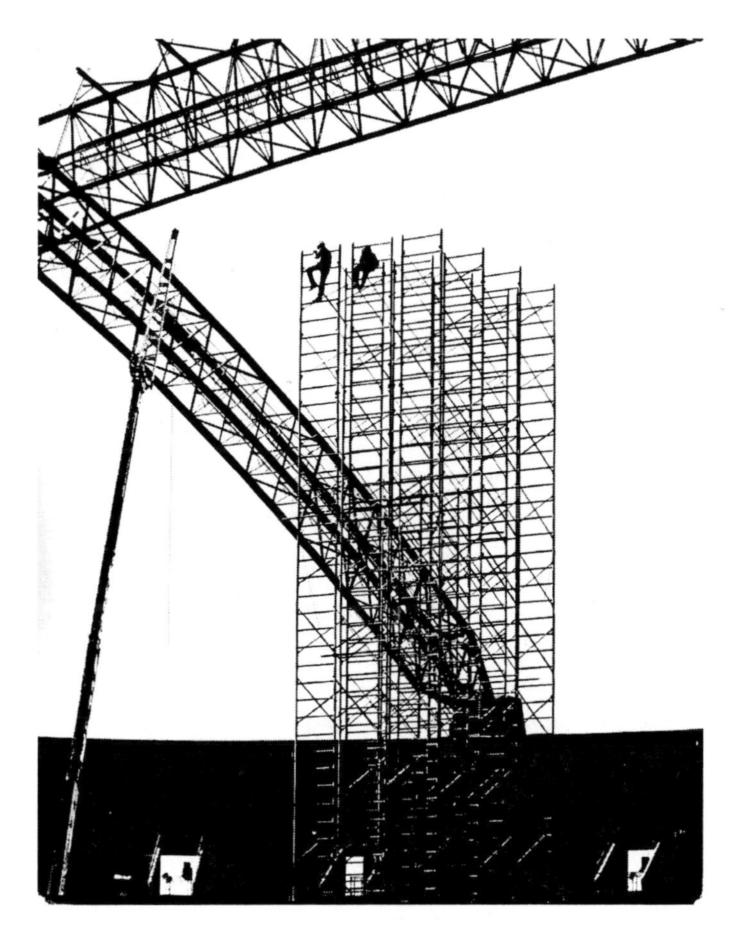

"Stadium Construction"

This big 16 by 20-inch black and white exhibition print, dry mounted on heavy poster board, is absolutely awesome to someone used to 4 x 6 color snapshots from the drug store. The starkness of the black and white and the incredible details are something never before seen, or better, felt. The tiniest hairline of black or white is crisp and sharp, and the image so realistic it transports the viewer into the scene. The tapering cable from the towering crane in "Stadium Construction" looks like the shadow of a spider's silk thread. Kilby lived in a world of photography unknown to the average person.

Jack traveled the world in his jobs, and for years carried his Hasselblad with him. The archive listing of his hundreds of negatives reads like a world tour—London, Mexico, Russia, Belgium, Japan, Switzerland, and many more that were shot in familiar locations, such as the Dallas zoo, and Fair Park, and of the family dog, Duchess, in the Kilby's front yard. His range of subjects is as broad as his range of locations. Jack's daughter Ann said, "I'm just struck when I look at the prints of his that I have, how many times he must have sat and watched people, and then took their pictures. Very close-up, too. It just seems so different from the way he came off without the camera."

Every professional photographer knows that taking the picture is only a small part in the complex process of making a great print. Jack had his photographic lab and darkroom equipment in the workroom that he shared with Barbara and her pottery-making studio. He had the equipment to develop his black and white films, and to print and enlarge the images. Jack trained his eye and his skills to "push" the contrast and tone of his enlargements, and to crop them to present what he felt was the perfect balance of light and dark, objects and space.

Jack's photographs were not of nature's grand events—the rainbows and the storms—but of the nuances of life. He froze the cigarette butt on the ground by the high heels, the reflection of a trombone in the shades of the musician, the glint in the eye of the old farmer with a face like a drought-cracked field. But that's way too arty-sounding for Jack and his work. His titles give a clue to his feelings of simplicity and strength—"Sweeper," "Iron," "Musician."

Photography is a lonesome experience: one man, one camera, one elusive image, and patience. Lots of patience. This very personal experience contrasted greatly with Jack's working world, which was filled with meetings, people, and impossible schedules. As a photographer, it was Jack and Jack alone who determined the outcome, so different from the production line in a factory. Ann said she thought he liked the quick response of the photographic arts compared to the snail-paced progress in his world of engineering. Jack could take a picture, massage it gently it to his exact liking, and have it propped up on an easel to admire in a matter of days, if not hours. Photography must have been an eminently satisfying and life-balancing avocation for Jack. He was good at it. Very good. And, oh yeah, he had fun at it, too.

"Speed"

Chapter Twenty

Jack makes sawdust

"A good hobby lasts seven years."
—Jack S. Kilby

JACK'S ARTISTIC SKILLS were exercised to their utmost in his photographic hobby—he must have used a delicate, almost spiritual, approach to bring out the exquisite images. And when he'd had enough of the exquisite stuff, Jack would squeeze down the narrow stairway to the garage below and change hobbies.

The row of fluorescent shop lights would come on, and eyes accustomed to the darkroom would squint as the room full of woodworking equipment came to life. Soon the civilized click of the Hasselblad's Synchro-Compur shutter would be forgotten in the shriek of a 10-inch table saw ripping the length of a seasoned hardwood plank. Now *that's* more like it! The nuance of one hobby balanced by the brute force of another. Jack was making a piece of furniture for someone in his family, and doing it with great enthusiasm and professional skill.

Jack's woodworking shop had a lot of tools. There was no doubt that Jack followed the Hobbyist's Creed: "When in doubt, buy another tool." His big industrial-grade table saw was the cornerstone power tool of a craftsman, and it was surrounded by planer, band saw, drill press, and joiner. The long wall of the garage/shop sported a big workbench, overhead cabinets, and tools on a wall rack. And don't forget the rows of glass jars with screws. Somewhere were stored stacks of *Fine Woodworking* magazine.

Ann and Janet don't have a list of all the furniture and cabinet work that their father built, but Ann summed it up with, "He built a lot of furniture, considering he was working at the time." As a sampling, a "room full of bookcases," a custom-fitted teak bedroom suite, cabinetry for a house, and one very special project—two walnut cradles for the first two granddaughters born a few months apart.

Ann remembers being asked to sketch what she wanted the cradle to look like. She did, and Jack took it from there, stopping only to build a steam-bending rig to put the curves in the staves. You can bet Jack had a special smile on his face as he thinned a piece of straight-grained five-quarters walnut in his planer.

Janet has a special fondness for the cradle, and pleasant memories of her daughters, Caitlan, Marcy, and Gwen, taking their appointed turns in it. Ann's special cradle also comforted her girls, Erica and Katrina. Janet also admits she was "pretty astounded that Dad loved playing with babies and making them laugh with his silly faces and sounds." She thought maybe making babies smile was even more fun for her father than making the cradles.

Chapter Twenty one

Recognition

The world recognizes Jack St. Clair Kilby

THIS IS A LIST OF AWARDS that Jack St. Clair Kilby received during his lifetime. An effort was made to find all the honors presented to Jack, but it is unlikely that a complete list exists, and certainly not here. Apologies are in order to the organizations that gave awards to Jack and are not listed below, and for any errors. Extreme Googling not withstanding, awards were undoubtedly overlooked, to the detriment of this book. This list includes only a sampling of the many honorary degrees he received.

Jack received awards, medals, and prizes; was inducted into elite organizations; given honorary degrees; and elected Fellow. He shook hands with presidents. If anyone ever received the whole nine yards of recognition, it was Jack, and one might think that this plethora of praise could cause a slight warpage of personality. Perhaps just a little—maybe a stifled yawn or and occasional ho-hum—when another award was announced. But this never happened to Jack Kilby. Seemingly impervious to acclaim, Jack, to the best of our knowledge, graciously accepted each honor, made a low-key speech that everyone applauded like crazy, and went back home exactly the same man he was when he left.

Jack's frequent speeches were a matter of some entertainment to his friends and family. Charles Phipps related that Jack had two "canned" speeches that he used: Speech A and Speech B. Speech A was the short speech, running about 8 minutes. Speech B, the "long" speech, rambled on endlessly for 12 to 14 minutes. Janet Kilby's comments about this long-standing source of amusement was that she understood that his short speech was "Thank you," and his long speech, "Thank you very much."

After Jack had received word that he was going to receive The Really Big One, the Nobel Prize, he attended a hastily assembled press conference. A reporter asked Jack what was the first thing he did after receiving the stunning news, and Jack replied, "I made coffee." Way to go, Jack!

Jack, as much as anyone in the industry, understood that progress in the highly competitive semiconductor business was the result of the work of many engineers and scientists, in the U.S. and around the world. He was quick to recognize and value their contributions. In accepting the Nobel Prize in Stockholm, Jack commented, "It's gratifying to see the committee recognize applied physics, since the award is typically given for basic research. I do think there's a symbiosis as the application of applied research often provides tools that then enhance the process of basic research. Certainly, the integrated circuit is a good example of that. Whether the research is applied or basic, we all 'stand upon the shoulders of giants,' as Isaac Newton said. I'm grateful to the innovative thinkers who came before me, and I admire the innovators who have followed."

Jack summed up his feelings nicely by quoting the laser Nobelist, Charles Townes, "It's like the beaver told the rabbit as they stared at the Hoover Dam. 'No, I didn't build it myself, but it's based on an idea of mine!'"

In later interviews, Jack commented on the contribution of industry pioneer, Robert Noyce, and said he had no doubt that if Bob had lived, he would have shared the Nobel Prize.

Awards and Honors

1966 Awarded Fellow grade, Institute of Electrical and Electronics Engineers (IEEE)

1966 Outstanding Achievement Award, Dallas Section IEEE

1966 Stuart Ballantine Medal, The Franklin Institute

1966 David Sarnoff Award, IEEE

1967 Elected to the National Academy of Engineers

1970 National Medal of Science, National Science Foundation, presented by President Nixon

1971 Alumni Honor Award, College of Engineering, Univ of Illinois

1973 Distinguished Alumni Award, Univ of Illinois

1974 Alumni Achievement Award, Univ of Illinois

1975 Vladimir Zworykin Award, National Academy of Engineering

1978 Co-recipient (with Robert Noyce) of the Cledo Brunetti Award, IEEE

1980 Consumer Electronics Award, IEEE

1982 Inducted into the National Inventors Hall of Fame, U.S. Patent Office

1982 Holley Medal, American Society of Mechanical Engineers (ASME)

1986 Medal of Honor, IEEE

1986 Distinguished Service Citation, College of Engineering, Univ of Wisconsin

1987 Patrick E. Haggerty Innovation Award, Texas Instruments Incorporated

1987 Coors American Ingenuity Award and Hall of Fame

1988 Official Texas Historical Marker, State of Texas

1988 Inducted into the Engineering and Science Hall of Fame, Dayton, Ohio

1989 Holley Medal presented by the ASME, jointly with Jerry Merryman and James Van Tassel, for the hand-held calculator

1989 Co-recipient (with Robert Noyce) of the Charles Stark Draper Prize, National Academy of Engineering

1990 National Medal of Technology, presented by President Geo. H. W. Bush

1993 Kyoto Prize for Technical Achievement, Japan

1994 Computer Society Pioneer Award, IEEE

1995 Robert N. Noyce Award, Semiconductor Industry Association

1999 Vladimir Karapetoff Award, Eta Kappa Nu, Electrical and Computer Engineering Honor Society

2000 The Harold Pender Award, Univ of Pennsylvania School of Engineering

2000 The Nobel Prize in Physics

Honorary Degrees

Rochester Institute of Technology – Doctor of Science

Yale University – Doctor of Science

Georgia Institute of Technology – Doctor of Philosophy

University of Miami – Doctorate

University of Illinois – Doctorate

University of Wisconsin – Doctorate

Chapter Twenty two

Kilby odds & ends

Leftovers, favorite stories, little-known facts, and other interesting and possibly shocking information about Jack Kilby

JACK WAS A PRIVATE MAN. That doesn't mean he was purposely secretive—he just didn't have much to say. Kind of like the title of this book—he was a man of few words. He might have heard a friend of mine's advice to his wife years ago when he told her, "Don't say Blah Blah Blah, when Blah will do."

He was totally transparent to the fact that others might be interested in his, say, photography, or that he built exquisitely crafted furniture in his woodworking shop in the garage. It never occurred to Jack to mention that he'd just won best-of-show for a photo he'd shot, processed, enlarged, cropped, printed, and entered in an international photographic salon.

But let's be honest about this "not much to say" trait and reputation. The engineers and scientists who worked with Jack in the development and engineering labs over the years at Texas Instruments will put up an argument about who's quiet and who's not. And Jack was not. Not, that is, if he was in a technical discussion. This was not the private Jack, but Jack the Engineer, doing engineering.

Since I find quotes "by the author" a bit much, forgive me as I lapse into occasional first-person story-telling.

Charley Clough, from a phone conversation

Charley was another long-time friend of Jack's. They went through the tough times of the early IC years at TI, with Charley on the marketing side and Jack on the technical. Such times make fast friends.

"Jack is a hard guy to describe. To me, he was my best friend. We just hung out together. And I don't remember any discussion that was beyond the routine. He was just very easy to be with, you know? And there were never any highlights or any big discussions. Yeah, I really miss the guy."

"...like so many Texans, Jack didn't like [Bill] Clinton. [Then] he got some award Clinton gave him about a year or two before he left the presidency, and he went up to get this award. Well, Jack came back, and I said, 'Hey Jack, that must have been pretty cool to accept that award from your old friend Bill Clinton,' joking, and he said, 'Charley, I was totally wrong about that guy. I've never met a guy like that. My god that guy, that's one of the most likeable men I ever met.'"

Jack's broad circle of friends

There is generally a strange barrier between engineers, back in the laboratory sweating over their soldering irons and oscilloscopes, and marketing people, ostensibly out peddling the engineers' precious creations with the help of martini-powered lunches and European spas. There seems to be little in common except the product itself, but Kilby ignored this dichotomy of professions and counted a number of marketing professionals among his best friends. Charley Clough is a case in point, as is Charles Phipps. Both of these senior marketing TIers worked closely with Jack in the early days of the integrated circuit and formed life-long friendships. Later, in the early 1980s, through the mutual friendship of Clough, Kevin McGarity, Senior Vice President of World-wide Sales and Marketing for TI at the time, became a fast friend of Jack's. It was Kevin and his wife Kathy who Kilby invited to go to Stockholm for the Nobel ceremonies. Jack just never understood how prejudices were supposed to work.

Shirley Sloat and her Kilby stories

Jack and other friends were at Shirley's house for dinner, and I was assisting her by wandering around and topping off people's wine glasses. I asked Shirley if she was ready for another glass, and she replied, "No, if I have two glasses of wine I get silly." Kilby, who was sitting nearby, chimed in, "If I drink a second glass, I get surly." This brought gales of laughter from those within earshot, obviously underwhelmed by the thought of Jack being surly.

Shirley, although from Michigan, had wholeheartedly accepted the Southern custom of black-eyed-peas-for-good-luck-on-New-Year's-Day tradition, and every year would take Jack a small bowl. Jack accepted her gift, with thanks, for several years, and then finally politely mentioned to Shirley that "more than one black-eyed pea is redundant." He would accept her kind gift, but let's not overdo the black-eyed peas was the message. Ever after, Shirley would put a couple of peas, no more, in a tiny plastic condiment container for Jack's New Year's Day good luck charm. The single black-eyed pea must have worked, since a short time later he received the Nobel Prize.

Texas A&M Symposium, March 2001

Shirley and I were delighted to be invited to the Texas A&M Symposium, which had been set up to honor Kilby's Nobel award. After the morning speakers, we adjourned to an adjacent hall to allow Jack to meet several of the top students and see their research posters. I had been watching one young man, dressed in his best, standing nervously in front of the easel with his project poster. I could see beads of sweat even from where I was, and he looked like he was on the verge of hyperventilating. Jack entered the room at his usual leisurely stroll, and the eyes of every student locked on him. He stopped at the student I'd been watching, stuck out his hand, and said, "Hi. I'm Jack Kilby." Only Jack would think he might need to introduce himself to the students. Only Jack had the total lack of ego. The good news is that student didn't faint, and as Jack left I could tell by the size of the smile on the student's face that he'd had the experience of a lifetime.

Diane Sawyer's famous question

Diane Sawyer did an extensive TV interview of Jack that was aired October 4, 1983. Near the end, she turned the conversation back to his current work on "a cheaper, better kind of solar energy." Diane looked at Jack's hieroglyph-filled blackboard, and pointing to a cluster of mysterious squiggles in one corner, asked Jack, "Is that what's up on the blackboard?" Jack answered without raising an eyebrow, "No, part of that is a map as to how to get to the guy who's fixing my car." To Diane's credit, she broke down laughing and concluded her interview with, "So much for smart questions!"

Jack reads

As a child, Jack read a lot, and Charles Phipps verified that Jack was still a voracious reader, especially on business trips. He would always have a couple of fiction paper-backs to pass the travel and hotel time. I was not aware of this, and during some conversation that Shirley Sloat and I were having with Jack, the surprising (to us) fact came out that Jack liked the Patrick O'Brian series of fictional British navy seafaring tales. Shirley and I had just discovered this delightful collection of books, and we agreed with Jack that they were the best sailing stories ever written; better even than the Horatio Hornblower series by C. S. Forester. O'Brian spent 30 years writing the 20 "linked" novels of Captain Jack Aubrey and his faithful sidekick, Stephen Maturin, physician, in the age of "fighting sail."

The next time I saw Jack, he handed me a grocery sack full of Patrick O'Brian stories—his collection of about 15 books. "And I don't want them back. I've read them." Shirley and I wallowed in these treasures for months, until Shirley had read them all, and I'd given up. Shirley discovered that there were a few more books that Jack didn't have, and bought them to fill out the collection.

This led to two further adventures. The first, we talked Jack into going to the movie "Master and Commander: the Far Side of the World" with us, to see Hollywood's Aubrey and Maturin in person. We enjoyed it greatly—the characters seemed like old friends.

A year or two later, we were talking with Ann Kilby's daughter, Erica, a budding writer, and found that she also liked Patrick O'Brian. Shirley gave her the complete set of O'Brian books. "And I don't want them back," I think I heard her say. The books had completed their proper journey.

Saturday Coffee with K & K

I was a latter-day participant in a Saturday morning ritual with Kilby and E. C. "Steve" Karnavas, known locally as "Coffee with K & K," beginning in 1990 or 1991. Steve had begun work at TI in 1952 after co-oping at SMU and receiving his mechanical engineering degree. One of his first full-time jobs was doing the mechanical design on the germanium crystal puller that put TI in the transistor business. He met Kilby shortly after Jack had come to work in 1958, and soon the Kilby family and the Karnavas family became best of friends. Steve later left TI to become the manager of Textool Products, which had no effect on the Saturday K & K coffee.

Every Saturday morning since the dawn of time, Kilby and Karnavas met at the Karnavas house for coffee, and I was honored to be invited to join them. The coffee was excellent, and we all took turns bringing doughnuts. What's not to like?

I soon learned a remarkable and important lesson from these get-togethers—there doesn't have to be continuous conversation while three people are drinking coffee. With Kilby and Karnavas, periods of silence, though at first seeming awkward to me, were soon shown to be perfectly natural to allow the sporadic interchanges to proceed at the proper pace.

These coffee times, which I participated in for a dozen years, give or take, were as informal as three engineers could make them, and that was plenty informal. Steve's wife, Margaret, was careful not to intrude, but she added a needed touch of culture to the gathering, and occasionally added needed doughnuts.

Margaret recently recalled to me the time she shared a nagging technical problem with the master of technical problems, Jack—she couldn't hear the front door bell when she was in her sewing room. Jack, ever brief, had the simple and foolproof solution—get a louder doorbell. This was followed by Jack's rare sly smile.

The most fun Saturday morning coffees were the rare "K & K Show and Tell" days. With Kilby, it was a one-sided event. First, he brought the highly significant "Kyoto Prize" that Japan had awarded him. Steve and I passed the gorgeous jeweled gold medal around and were mightily impressed. Jack didn't exactly play "show & tell" fair. Seven years later he brought in his Nobel Prize medal for us to see. It, too, was gorgeous. We got to play with it. I don't remember what Steve or I brought that day.

Ed with Nobel

The only time I can recall Jack saying anything that remotely reflected on his prominence was when he, out of the clear blue at a Saturday coffee, mentioned that he'd gotten a call from Texas Instruments, and was told they were going to name a new research and development building after him. He seemed almost embarrassed to have brought it up, but he also seemed quite pleased about the matter, as he should have been. Later, as the Kilby Center neared completion, it was draped with a most appropriate gigantic banner: "The Chip that Jack Built Changed the World."

The Karnavas house was conveniently close to Jack's personal office where he went on Saturday after getting his caffeine and conversation fix. It was his time to answer his weekly mail. Jack's office was cluttered as engineers are wont, and the ashtray on his desk frequently in need of emptying. But large windows on the end of the room opened onto outdoor greenery, making his office a pleasantly informal place. Piles of papers and magazines topped the bookcases, and the coffee maker and cups were a bit spotted and stained. But Jack wasn't really messy, just unkempt in a casual way. The great majority of the people who would be visiting Jack would notice nothing unusual about the surroundings. Engineers, as a rule, are more interested in engineering than decor.

Kilby would sit alone in his office and with great patience open and answer his mail, mostly from strangers, and mostly from people wanting Jack's autograph. Jack had once mentioned at the coffee that several people had requested copies of his famous notebook sketch of the first integrated circuit. Not with a copy machine, mind you, but with his pen on a fresh piece of paper. A fresh piece of signed and dated paper, of course. And Jack obliged without batting an eye. Jack took his stardom seriously, and always courteously answered phone calls and letters without complaining. Well, he did complain a little about making the drawings.

After Jack received the Nobel Prize, he made a surprising request at one of our coffees. I had made a series of "engineer's note cards" for my own use, the fronts decorated with drawings of theodolites, cutaways of aircraft engines, Tesla coils, and the like, and Jack somewhat reluctantly asked if I could make some for him. If it wouldn't be too much trouble, four or five dozen? He said he needed to write a lot of thank-you notes and that he liked the décor of the cards. Of course it wasn't too much trouble, and Jack soon sent out an avalanche of thank-you cards. A month or so later, he asked, still reluctantly, for a couple of dozen more. After Jack died in 2005, the IEEE *Spectrum* magazine editor (as I recall) mentioned Jack's death in his column, and commented on the fact that Jack had sent him a personal note card after the Nobel ceremony, thanking him for his article in their magazine about him. The editor said that it so typified Jack—considerate and humble to a fault.

What serious matters did Nobelists talk about in their time off-camera? (Not that Jack ever had to get off camera because he was never on it.) One coffee-time conversation that Jack initiated was about the new rolls of self-adhesive stamps from the post office. Jack contended that the stamps were on the rolls upside down, and when a normal person pulled out the strip of stamps to peel one off, it would be the wrong way around to put on the letter. This spell-binding engineering discussion went on for some minutes but no satisfying conclusion was ever reached. A close second for fascinating topics was recalled by Shirley Sloat from one of her rare K & K attendances, "We talked about how to keep silverware from tarnishing, and Jack was pretty knowledgeable."

How would I sum Jack up? Just thinking about it brings a cloud of adjectives and adverbs circling over my head. But I feel that Jack's succinct manner of writing and speaking should be the key for his description. Jack Kilby was a good guy, and I, too, miss him.

Jack will remain in my heart, and in the hearts of those who knew him, not primarily as a great inventor and Nobelist, but as a man who knew what was right and never wavered and what was wrong and never succumbed.

K & K

Patents

Patent	Title	Issued	Inventor(s)	Filed	Assign
2,637,777	Electrical Network Having Distributed Capacitance	5/5/53	Kilby, Khouri	2/27/50	Globe-Union
2,759,854	Method of Manufacturing Capacitors	8/21/56	Kilby	6/20/51	Globe-Union
2,762,001	Fused Junction Transistor Assemblies	9/4/56	Kilby	3/23/55	Globe-Union
2,823,262	Telephone Answering Device	2/11/58	Kilby, Youngbeck	10/21/53	H.A. Milhaupt Inc.
2,841,508	Electrical Circuit Elements	7/1/58	Roup, Kilby	5/27/55	Globe-Union
2,883,499	Resistance Trimmer	4/21/59	Kilby, Khouri	11/24/53	Globe-Union
2,892,130	Plug-in Circuits	6/23/59	Kilby	12/16/53	Globe-Union
2,945,163	Component Mounting for Printed Circuit	7/12/60	Kilby, et al	1/10/55	Globe-Union
3,052,822	Modular Electrical Units and Assemblies Thereof	9/4/62	Kilby	5/28/58	Globe-Union
3,069,261	Method of Making Porous Metal Bodies	12/18/62	Kilby, Williamsen	10/25/57	Globe-Union
3,072,832	Semiconductor Structure Fabrication	1/8/63	Kilby	5/6/59	TI

Patent	Title	Issued	Inventor(s)	Filed	Assign
3,115,581	Miniature Semiconductor Integrated Circuit	12/13/63	Kilby	5/6/59	TI
3,134,049	Modular Electrical Units and Assemblies Thereof	5/19/64	Kilby	5/13/58	Globe-Union
3,138,721	Miniature Semiconductor Network Diode and Gate	6/23/64	Kilby	5/6/59	TI
3,138,743	Miniaturized Electronic Circuits	6/23/64	Kilby	2/6/59	TI
3,138,744	Miniaturized Self-Contained Circuit Modules and Method of Fabrication	6/23/64	Kilby	5/6/59	TI
3,211,972	Semiconductor Networks	10/12/65	Kilby, Lathrop	6/24/64	TI
3,222,610	Low Frequency Amplifier Employing Field Effect Device	12/7/65	Evans, Kilby	5/2/69	TI
3,261,081	Method of Making Miniaturized Electronic Circuits	7/19/66	Kilby	2/6/59	TI
3,304,429	Electrical Chopper Comprising Photosensitive Transistors and Light Emissive Diode	2/14/67	Kilby, Bonin	11/29/63	TI
3,304,431	Photosensitive Transistor Chopper Using Light Emissive Diode	2/14/67	Kilby, Biard, Bonin, Pittman	11/29/63	TI

Patent	Title	Issued	Inventor(s)	Filed	Assign
3,313,986	Interconnecting Miniature Circuit Modules	4/11/67	Kilby	12/23/65	TI
3,340,406	Integrated Semiconductive Circuit Structure	9/5/67	Kilby	1/24/67	TI
3,350,760	Capacitor for Miniature Electronic Circuits of the Like	11/7/67	Kilby	3/16/64	TI
3,364,397	Semiconductor Network Inverter Circuit	1/16/68	Kilby	1/13/67	TI
3,386,187	Teaching Machine	6/4/68	Kilby	2/28/66	TI
3,411,051	Transistor with an Isolated Region Having P-N Junction Extending from the Isolated Wall to a surface	11/12/68	Kilby	12/29/64	TI
3,413,480	Electro-Optical transistor Switching Device	11/26/68	Biard, Bonin, Kilby, Pittman	11/29/63	TI
3,434,015	Capacitor for Miniature Electronic Circuits or the Like	3/18/69	Kilby	2/17/67	TI
3,435,516	Semiconductor Structure Fabrication	4/1/69	Kilby	1/13/67	TI
3,436,604	Complex Integrated Circuit Array and Method for Fabricating Same	4/1/69	Hyltin, Kilby, Luecke, Toombs	4/25/66	TI
3,484,534	Multilead Package for a Multilead Electrical Device	12/16/69	Kilby, Toombs, Van Tassel	7/29/66	TI

Patent	Title	Issued	Inventor(s)	Filed	Assign
3,496,333	Thermal Printer	2/17/70	Alexander, Emmons, Kilby	9/26/69	TI
3,598,664	High Frequency Transistor & Process for Fabricating Same	8/10/71	Kilby	12/6/67	TI
3,643,138	Semiconductor Device	2/15/72	Kilby	1/29/62	TI
3,643,232	Large-scale Integration of Electronic Systems in Microminiature Form	2/15/72	Kilby	6/5/67	TI
3,688,396	Circuit Board Process	9/5/72	Kilby, Van Tassel	10/13/69	TI
3,696,411	Keyboard Encoder	10/3/72	Kilby, Van Tassel	11/12/70	TI
3,698,082	Complex Circuit Array Method	10/17/72	Hyltin, Kilby, Luecke, Toombs	2/25/71	TI
3,711,626	Circuit Board	1/16/73	Kilby, Van Tassel	9/29/69	TI
3,777,067	System for Disabling Incoming Telephone Calls	12/4/73	Kilby	12/30/71	Kilby
3,781,866	Binary Encoding Switch	12/25/73	Kilby	6/18/71	Kilby
3,784,721	System for Screening Telephone Calls	1/8/74	Kilby	11/22/71	Kilby
3,793,487	System for Screening Telephone Calls	2/19/74	Kilby	6/26/72	Kilby
3,819,921	Miniature Electronic Calculator	6/25/74	Kilby, Merryman, Van Tassel	12/21/72	TI

Patent	Title	Issued	Inventor(s)	Filed	Assign
3,835,530	Method of Making Semiconductor Devices	9/17/74	Kilby	9/22/71	TI
3,898,450	Reliable Flashlight	8/5/75	Kilby	11/1/73	TI
3,920,979	Electronic Check Writer	11/18/75	Kilby, Schweitzer, McCrady	10/19/73	Kilby
3,944,724	Paging System with Selectively Actuable Pocket Printers	3/16/76	Kilby, Schweitzer, McCrady	5/18/72	TI
3,955,354	Display for Electronic Clocks and Watches	5/11/76	Kilby, Schweitzer	2/11/74	Kilby
3,979,757	Electrostatic Display System with Toner Applied to Head	9/7/76	Kilby, Lathrop	1/15/75	Kilby, Lathrop
3,979,758	Electrostatic Head with Toner Attracting Plates	9/7/76	Kilby, Lathrop	1/20/75	Kilby, Lathrop
4,001,947	Teaching System	1/11/77	Kilby	8/16/73	Kilby
4,021,323	Solar Energy Conversion	5/3/77	Kilby, Lathrop, Porter	7/28/75	TI
4,042,948	Integrated Circuit Isolation with Mesas and/or Insulating Substrate	8/16/77	Kilby	5/20/74	TI
4,090,059	Thermal Recording Head for Printer	5/16/78	Kilby, Schweitzer, McCrady	10/20/75	TI
4,100,051	Light Energy Conversion	7/11/78	Kilby, Lathrop, Porter	12/2/76	Kilby

Patent	Title	Issued	Inventor(s)	Filed	Assign
4,136,436	Light Energy Conversion	1/30/79	Kilby, Lathrop, Porter	12/2/76	TI
4,173,494	Glass Support Light Energy Converter	11/6/79	Kilby, Johnson, Lathrop, McFerren, Myers	2/14/77	Kilby
4,188,177	System for Fabrication of Semiconductor Bodies	2/12/80	Kilby, McKee, Porter	2/7/77	TI
4,270,263	Glass Support Light Energy Converter	6/2/81	Johnson, Kilby, Lathrop, McFerren, Myers	8/10/78	TI
4,322,379	Fabrication Process for Semiconductor Bodie	3/30/82	Kilby, McKee, Porter	4/2/79	TI
5,611,884	Flip Chip Silicone Pressure Sensitive Conductive Adhesive	3/18/97	Bearinger, Camilletts, Kilby, Haluska, Michael	12/11/95	Dow Corning

Sources

The sources of information for the early chapters of this book are scarce. Two of them require special mention. The first sources are Pam Karnavas's irreplaceable personal notes of her interviews with Jack and Jane Kilby which she so generously shared with me. The other are the personal files of Jack donated by his family and now archived in the DeGolyer Library at Southern Methodist University in Dallas. Dr. Anne Peterson, Curator of Photographs of the DeGolyer, was more than kind in helping me access the files and photographs in her care.

Special thanks to the University of Illinois archives for their prompt assistance and kind permission to use the material about Jack's college days as described in chapters seven and nine.

Special thanks also to the Southern Illinois University Press for their permission to quote from Oliver Caldwell's *A Secret War: Americans in China 1944-1945,* found in chapter eight.

Chapters 1 through 7

Articles

Lifetimes newspaper section, unknown source, March 2005, interview with Jane Kilby.

Norberg, Arthur L. "An Interview with Jack S. Kilby," Charles Babbage Institute interview on June 21, 1984, Dallas, Texas. OH74. Center for the History of Information Processing, University of Minnesota, Minnesota.

Internet web sites, as of May, 2008

www.kshs.org Kansas State Historical Society

www.jackkilby.com

QRZ.com (W9DKI) – Amateur radio operators
Nobelprize.org/nobel_prizes/physics/laureates/2000/Kilby-autobio.html – Kilby autobiography
Wikipedia.org – Cities, towns
www.pinetreeweb.com/1937-nj1 – Boy Scout National Jamboree

Additional sources

Pamela M. Karnavas interview notes, 1989–1990.
University of Illinois archives.
DeGolyer Library, Southern Methodist University, Dallas, Texas, A2006.0032.

Chapter 8

Books

Campbell, Jeff C. *Vignettes.* Houston, Texas: the author, 2003. Descriptions of truck driving on the Burma Road.
Caldwell, Oliver J. *A Secret War: Americans in China 1944–1945.* Carbondale and Edwardsville, Illinois: Southern Illinois University Press, 1972. Two paragraphs quoted with their kind permission.

Articles

Lifetimes newspaper section, unknown source, March 2005, interview with Jane Kilby.
Phipps, Charles. "Jack S. Kilby – 8 November 1923–20 June 2005." *Proceedings of the American Philosophical Society*, Vol. 151, No. 4, December 2007. (Biographical memoirs)
Norberg, Arthur L. "An Interview with Jack S. Kilby."

Additional sources

Pamela M. Karnavas interview notes, 1989–1990.
DeGolyer Library, Southern Methodist University, Dallas, Texas, A2006.0032.

Personal conversations

Charles Phipps, May 5, 2008 – military experiences related by Kilby.

Harvey Cragon, April 21, 2008 – military experiences related by Kilby.

Chapters 9 through 11

Articles

Norberg, Arthur L. "An Interview with Jack S. Kilby."

Additional sources

DeGolyer Library, Southern Methodist University, Dallas, Texas, A2006.0032.

Chapters 12 through 16

Books

Pirtle, Caleb III. *Engineering the World – Stories from the First 75 Years of Texas Instruments,* Dallas, Texas: Southern Methodist University Press, 2005.

Articles

Kilby, Jack S. "Invention of the Integrated Circuit." *IEEE Transactions on Electron Devices,* Vol. ED-23, No. 7, July 1976.

Kilby obituary, *Dallas Morning News,* June 22, 2005.

Kilby obituary, *New York Times,* June 22, 2005.

Norberg, Arthur L. "An Interview with Jack S. Kilby."

Additional sources

DeGolyer Library, Southern Methodist University, Dallas, Texas, A2006.0032.

Pamela M. Karnavas interview notes, 1989–1990.

Kilby TI engineering notebook, courtesy Texas Instruments.

Kilby post-TI engineering notebooks, courtesy Ann and Janet Kilby.
Early IC, calculator information, solar energy information, courtesy Texas Instruments.

Personal conversations

Charles Phipps April 18, 2008 – early IC days.
Jerry Merryman, April 27, 2008 – calculator design information.
Pete Johnson April 28, 2008 – solar energy project.

Chapters 17 through 20

Internet web sites

www.dallascad.org – Dallas Central Appraisal District.

Additional sources

DeGolyer Library, Southern Methodist University, Dallas, Texas, A2006.0032.
Pamela M. Karnavas interview notes, 1989–1990.
Dallas Camera Club archives, via Ken Zapp.

Conversations and emails

Ann Kilby and Janet St. Clair Kilby – the majority of the family information, used with their kind permission.
Steve and Margaret Karnavas
Ken Zapp – Dallas Camera Club information.
Charles Phipps
Winky Waugh – photography information.
Charley Clough

Chapter 21

Most of the awards and honors information courtesy Texas Instruments.
www.google.com

Chapter 22

Personal

Ed Millis – Saturday coffee, other.

Conversations

Charley Clough
E. C. "Steve" and Margaret Karnavas
Charles Phipps
Shirley Sloat
Kevin McGarity

Additional sources

Video tape, Diane Sawyer interview October 4, 1984, courtesy Ann and Janet Kilby.

Images

Frontispiece

"An Engineer… " – courtesy of Texas Instruments

Chapter 2

Boy Scout – DeGolyer Library, Southern Methodist University, Dallas, Texas, Ag2006.0010 (photographs).

Chapter 3

Youthful Kilby – courtesy Ann and Janet Kilby.

Chapter 5

Kilby and supercharged Model A Ford – courtesy Ann and Janet Kilby.

Kilby and restored Model A Ford – Ed Millis.

Chapter 8

"Kilby in the doorway…" – DeGolyer Library, Southern Methodist University, Dallas, Texas, Ag2006.0010 (photographs).

"How did Jack get a TI jeep?" – DeGolyer Library, Southern Methodist University, Dallas, Texas, Ag2006.0010 (photographs).

"Kilby dines…" – courtesy of Ann and Janet Kilby.

Chapter 12

Kilby's TI notebook scans, 13 each – courtesy of Texas Instruments.

Chapter 13

Solid Circuits and match head – courtesy Texas Instruments.

Chapter 14

Kilby and calculator – DeGolyer Library, Southern Methodist University, Dallas, Texas, Ag2006.0010 (photographs).

Chapter 17

Kilby family on vacation – DeGolyer Library, Southern Methodist University, Dallas, Texas, Ag2006.0010 (photographs).

Chapter 19

"Stadium Construction," by Kilby – Ed Millis.
"Speed," by Kilby – Ed Millis.

Chapter 22

"Ed with Nobel" – Ed Millis.
"K & K" – courtesy either Margaret Karnavas, Shirley Sloat, or Ed Millis.

Index